Textbook of Human Nutrition and Dietetics

Textbook of Human Nutrition and Dietetics

– Authors –

Dr. S.C. Mozumder

M.B.B.S.
Medical Practitioner,
Asansol, India.

and

Dr. Bibhas Chandra Mazumdar

M.Sc. Agri., Ph.D.
Former, Professor,
Institute of Agricultural Sciences,
Calcutta University,
Kolkata, India.

2025

Daya Publishing House®
A Division of
Astral International Pvt. Ltd.
New Delhi – 110002

ISBN 9789359198538

Published by : **Daya Publishing House®**
A Division of
Astral International Pvt. Ltd.
– ISO 9001:2015 Certified Company –
4736/23, Ansari Road, Darya Ganj
New Delhi-110 002
Ph. 011-43549197, 8130496929
e-mail: info@astralint.com
Website: www.astralint.com

Dedicated to

Ms Sharmishtha Mozumder

— a lady of many rare qualities

Preface

Functional processes of the body systems in a living being is an inseparable part to comprehend the transcentral functioning of the Mother Nature. To keep the set of activities of the diverse organs of the body in motion, provision of specific agents from exogenous source is however, of dire necessity and ostensibly, these are made available in the body in the form of food and drink. The health science has made it known that growth of an individual from its birth till last breath is not possible to be achieved unless the body regularly gets food and drink.

Necessity of the victuals to sustain life had come to the knowledge even by the first form of life that was born on this planet. The primitive human beings were also indubious to concede that food and drink were indispensable items not only to satisfy their hunger, thirst and satiety but also to carry on the daily activities effectively, although they were ignorant of how these agents were utilized in their body systems and render benefit. It was Antoine Lavoisier, the eminent scientist of the later part of 1770s who had first brought to light that the living body could get heat energy when the food taken would get oxidized. Since that time, series of studies had been undertaken by the scientists and eventually, the new science known as *nutrition* had taken its birth. *The* complementary discipline, *dietetics* had also emerged with the objective to peruse the role of diet in providing nutrition to meet the requirement of a particular person.

It was earlier believed that food and drink taken would undergo chemical transformation in the stomach and thereby, provide nourishment and energy. Extensive studies done thereafter however, unravelled that apart from energizing the body, plenty of other benefits could also be rendered by them. It was learnt that the alimental victuals were in fact, adequately equipped with numerous chemical substances and these were responsible to bring about salubrity. These chemical agents which make provision of nutrition were termed as *nutrients* and were known to be of *six* categories. It is now well-founded that the body gets energy, accomplishes growth, fights against diseases and maintains health on availability of the nutrients of various kinds.

Adequate importance on nutrition had perhaps missed attention in the area of health science till the nineteenth and the early part of the twentieth centuries. The physicians used to advise intake of wholesome food to the patients of those days as adjunct of medicine, in abiotrophy or emaciation of unknown cause. With passage of time, the health scientists however, took stock of the importance of nutrition with heedful attention and took up studies to derive knowledge on the implicit effect of nutrition on the physiological mechanisms. The precise role of nutrition in the daily diet of every person whether morbid or not had been confirmed. Nutrition and dietetics have assumed now separate sciences rather than a step-child of medicine. Accordingly, degrees and diplomas are offered by the universities and institutions on this vital and inseparable stream of health science.

The principal author of the present title in his five decades of medical practice has long thought of bringing out a textbook on human nutrition and dietetics for the Indian students incorporating in particular, his personal experiences. Jointly with the co-author, who is a scientist in the area of agricultural, horticultural and food sciences, the text in now possible to be released. For physical incapacity, the principal author had not been able to write by himself many portions of the textual matter and hence, he had to dictate those portions to the co-author. The co-author took great pain to rewrite and transliterate those parts and the principal author gratefully acknowledges the co-author for this effort.

In preparing the title, the authors have consulted a number of books, literary works and also the scholarly articles appeared in googles and electronic media, magazines, newspapers, leaflets *etc.* and they gratefully acknowledge these brilliant presentations. They acknowledge with sincerest gratitude the following texts in particular, the materials presented in which have been heedfully considered by them and some facts presented in these highly informative titles have been incorporated in the present monograph.

(1). Human Physiology, Vol. 1 (2004) by C. C. Chatterjee (Publisher, Medical Allied Agency, Kolkata - 700 009). (2). Essentials of Medical Physiology (2005) by A.B.S. Mahapatra (Publisher, Current Book International, 60, Lenin Sarani, Kolkata - 700 013). (3). Concise Medical Physiology (2004) by S. K. Choudhury (Publisher, Central Book Agency, Kolkata - 700 009). (4). Essentials of Food and Nutrition, Vol. 1 by P. Swaminathan (Publisher, Ganesh and Co., Chennai - 600 017). (5). Biochemistry (2004) by L. Veerakumari (Publisher, MJP Publishers, Chennai - 600 005). (6). Nutritive Value of Indian Foods by Gopalan *et al.* (2004). (Revised and updated by B. S. Narasinga Rao *et al.*) (Publisher, National Institute of Nutrition, under the I.C.M.R., Hyderabad - 500 007).

For correcting proof, arranging materials, modifying the language in some parts to present in more easily understandable form and other assistance, the authors gratefully acknowledge Mrs. Purabi Mazumdar, M.A., B.Ed., Sng. Pvr. Grateful acknowledgement is also due to two young scientists, Dr. Kaushiki Mazumdar, M.Sc., Ph.D. and Dr. N.K. Dutta, M.Sc., Ph.D., both in the pharmaceutical Technology Department of the Jadavpur University, Kolkata for providing many valued informations on the microbiological aspects of human nutrition and dietetics.

The title has been attempted to aim at meeting the requirements of the under-graduate and post-graduate students of human nutrition and dietietics, nursing, pharmaceutics, dental, ayurvedic, homoeopathic and unani sciences, besides the students of M.B.B.S. and M.D. in Nutrition, Physiology and Medicine.

As far as possible, the authors have used simple language and avoided less commonly used words for easy comprehension of the text to the readers. Eventually, the authors will be amply rewarded if the title is able to provide any benefit to them. Lastly, they invite informative and valued suggestions from the readers for enrichment and enlargement of the monograph in future.

S. C. Mozumder

Bibhas Chandra Mazumdar

Contents

Chapter 1

Basic Aspects of Nutrition

Greatest asset of human beings or any living organism is health. Its worth is not tantamount to wealth or any of the faculty acquired or inherited by an individual. Riches or resources can afford joy, delight and pleasure to a person but not happiness, bliss and peace of mind, if he or she is physically or mentally infirm or in ill health since birth or during passing of days in life, which robs his or her ability to perform activities with adroitness as is done by a healthy individual.

Health, the other name of which is good health has been viewed and defined by people of different conceptualism in different ways. The World Health Organization (WHO) of the United Nations has defined health as "a state of complete physical, mental and social well-being and not merely the absence of disease or infirmity". The definition is though accepted by many but has not been fully supported by a large body of health scientists for they find limited usefulness as regard implementation of the proposition. It has even been put forward by a section of health professionals that many persons are there who apparently enjoy a state of well-being, although on critical scrutiny they might be considered to be unhealthy in some form or other.

Whatever may be the connotations, it is candidly perspicous that good health of a person should be such that, it remains *free from any abnormality in a person in respect of physical or mental state.* Undesirable deviation from the normal functioning of the body occurs in fact, for a number of reasons and a pertinent one could be said to be *improper nourishment.* This is well-suggestive of the fact that while faulty or incorrect nourishment brings about unhealthy condition of various types, desirable and idealistic nutrition stand in the way against malfunctioning of various organs and systems in the body, besides defending against many pathogenic infections and physiological disorders.

1.1. Definition

In a broad sense, *nutrition* (Latin, *nutritio* = nourish) refers to the process of taking health-giving, *i.e.*, nourishing food (= nutriments) and their proper use in the body in regard to growth, metabolism and repair of body tissues. An analysis of the statement signifies that the process of nutrition involves two steps, which are firstly, what type of food is consumed by a person and secondly, how the consumed food benefits the body to sustain growth, carry on daily activities and maintain health.

The earlier health scientists conceived nutrition as a branch of medicine. However, in course time, the concept of nutrition has been greatly modified and much higher emphasis has been given to the daily diet of a person. At the present time, the health scientists have defined nutrition as the *sum total of the processes involved in regard to taking in and ultization of food substances by which, growth, repair and maintenance of the body are accomplished.* In a clearer manner, it has been stated as the "overall process by which cellular organelles, cells, tissues, organs, systems and the body as a whole tend to obtain necessary substances from food as sources of energy and to maintain structural and functional integrity".

A scrutiny of the above definition makes it apparent that nutrition in the body systems in fact, continues in several *steps* or *stages*. Sequentially, these steps are as follows: *Ingestion* of the food is the first step and this is followed by other steps, *i.e., digestion, absorption, transport* and *assimilation* which take place inside the body and the last step is *excretion,* which involves removal of the unwanted part of the food from the body that remains undigested or unutilized.

Such a definition of nutrition is also not fully accepted by all the nutrition scientists in the world and according to them, there are other conditions also besides these, which should be duly considered. It should also be remembered that nutrients do not always come from food, *i.e.*, food taken by mouth is not the only source of nutrition in the body. There is also a system by which the nutrients, *i.e.*, agents which provide nourishment may be directly injected into the blood stream in the form of injectable, *i.e.*, solution form. Again, nutrients may be delivered directly into the colon also. These are called *critical care nutrition* and has been discussed later.

It should not be forgotten in this context that many of the end-products that are accumulated in the body due to various metabolic processes are made utilized by the body as nutritive substances in various organs. But their role has not much significance.

1.2. History

Importance of nutrition to render beneficial effect on health of the living beings has not been suddenly discovered by the human beings at the present time. In fact, it has a long history since pre-historic times. Mention may be made to the fact that first evidence of the importance of food to relieve pain has been obtained from the *carving on the stone tablet in Babylon* as early as 2500 BC. Incidence of scurvy disease which was although extensively studied much later had first been known in the Ebers Papyrus in 1500 BC. In accordance with reactions that take place in the body, ancient people of China and India classified foods into *hot* or *cold* groups. Experiments on nutrition on the human beings had first been recorded in the Bible's Book of Daniel. Existence of substances in food which were beneficial to the health of human beings had been suggested by Anaxagoras at around 475 BC who termed them as *homeomerics* to denote generative components. Hippocrates at around 400 BC went on to comment, "Let food be your medicine and medicine be your food. "Classifying foods as *strong* and *weak* groups was done by the ancient Roman physician, Aulus Celsus Galen, who had also paid much attention on food that had been beneficial to the human beings.

Parcelsus in the 1500s and the eminent artist and scientist, Leonardo de Vinci had brought forward some newer aspects on nutritive value of foods, while criticizing the earlier propositions put forward by other workers. Experiments on scientific aspect of nutrition by James Lind in 1747, the surgeon of the British Navy led to the *remarkable discovery* of vitamin C to prevent and cure the scurvy disease of the British sailors. Antoine Lavoisier at around 1770 AD had demonstrated that heat in the body could be obtained by oxidation of food. The necessity of calcium for growth of the fowls had been put forward by George Fordyce in 1970.

Study of William Prout is worth mentioning, who in the year, 1827 had first divided foods under the groups of *carbohydrate, protein* and *fat*. Beriberi disease of the Japanese soldiers had drawn attention of Kanechiro Takaki who put forward the cause of that as due to high consumption of white rice. Importance of the chemical element, *iodine* in thyroid functioning was pointed out by Eugen Baumann in 1896. Measuring *expenditure of energy in terms of calorie* was put forward by the scientists, Carl von Volt and Max Rubber on carrying out independent studies. The thought provoking statement of Frederick Hopkins in 1906 that some unsuspected dietetic factors other than protein, calorie and minerals had their necessity to get rid of deficiency diseases shook attention of many scientists.

The terms, *vitamin* was introduced by Casimir Funk in 1912 by combining the words, *vital* and *amine* as it was thought that deficiency of some amino acids were responsible for some diseases like beriberi, pellagra *etc.* However, later it was known that *all vitamins were not actually amino acids.* Elmer McCollum and Marguerite Davis in 1913 and 1915 were the pioneer workers to discover vitamin A and some of those of the B group. Requirement of very small amount of *copper for absorption of iron* in the body had been pointed out by Hart in 1925. For synthesis of vitamin D in 1927, Adolf Otto R. Windaus was awarded the prestigious Nobel Prize. The role of essential *amino acids* on *synthesis of protein molecules* was propounded by William Cumming Rose in the 1930s. Credit goes to Erhard Fernholz for discovery of the chemical structure of the vitamin E. The first *Recommended Dietary Allowances* were established in the United Kingdom in 1941 and the United States Department of Agriculture introduced the *Food Guide Pyramid* in the year, 1992, which is of high practical usefulness.

Under Indian condition, scientific researches on food and nutrition had first been taken up by the eminent chemist, Acharyya Prafulla Chandra Roy (1861 – 1844), the Palit Professor of Chemistry of the Calcutta University in the later part of 1800 and early part of 1900. Apart from his research studies in Chemistry on mercurous nitrate and other areas, this scientist, patriot and social reformer became highly interested to undertake research and development on food science, nutrition and dietetics, on observing the poor food habit of the common people of Bengal province and other parts of eastern India. Acharyya Roy had primarily centered attention on low-cost foods that could be *affordable and traditionally consumed by the poor people* and took up the mission on bringing awareness of *nutrition* to the people. He had to his credit a number of books on food and nutrition for the common people.

The name of another scientist, Dr. Chuni Lal Bose is also worth mentioning who at around the same time undertook series of studies on human nutrition and dietetics and authored a number of books for the common people besides publishing many research articles. Dr. Bose, the then Professor in the Medical College in Calcutta proposed the concept of *balanced diets* for the Indian people. For his contribution in the areas of chemistry, food science, nutrition and medicine, the British Govt. conferred on him a special honour (Rai Bahadur) in 1898. In subsequent years, Dr. Amiya Kumar Bose, the Professor of Obstetrics and Gynaecology of the National Medical College, Calcutta devoted serious attention on research and development on nutrition. His researches on the role of vitamin C during pregnancy and on other areas are significant contribution. Dr. Bose established the Society of Nutrition and founded the journal, *Applied Nutrition* with Prof. (Dr.) Bibhas Chandra Mazumdar as the Publication Secretary.

However, India takes the pride on scientific researches on food, nutrition and dietetics from its highly prestigious centre, which is National Institute of Nutrition at Hyderabad. The Institute which comes under the control of the Indian Council of Medical Research of the Govt. of India is keenly devoted to the research and development on food science, nutrition and dietetics. Apart from undertaking critical and sophisticated researches, the Institute is also accredited to impart education and confer post-graduate and under-graduate degrees and diplomas on various aspects of nutrition and health science. The Institute was founded by Sir Robert McCarrison as early as 1918 in a single room laboratory at Coonoor for the purpose of studying the Beriberi disease. At present, it is considered as one of the biggest Food and Nutrition Institutes in the world which is housed with many departments and centres.

The scientists at this Institute are devoted to carry on *intensive researches* on various aspects in the areas on *nutrition, food* and *dietetics* as well as *on related fields.* Among the volumes of invaluable studies that have been undertaken by the scientists at this Institute, some may be mentioned as nutritional profiles of different communities, high risk pregnancy in the rural areas, pancreatic studies, enrichment of iodized salts *etc.* For remarkable contribution, the World Health Organization

(WHO) and the Food and Agricultural Organization (FAO) of the United Nations *had recognized this Institute as the Centre of Excellence.* Mention may also be made in this context that advanced level of researches on food, nutrition and dietetics and related areas are being carried out now in many universities and Institutes in India. The country has in fact, made a promising stride on research and development in the areas on food, nutrition and dietetics. To impart education on nutrition and dietetics, many universities and colleges in India have now opened up courses at the post-graduate, under-graduate and diploma levels.

1.3. Nutrients

1.3.1. Definition

Nutrition takes place in the body by the intake of solid or liquid foods that are of high food value. These foods which are known as nutritive (= nutritious, alimental) foods are known to contain some chemical substances, which actually bring about nourishment (= alimentation) in the body. These nourishing substances obtained from the foods are termed as *nutrients* or *nutriments* and are also recognized as nutritional components. The nutrients are therefore, *chemical substances present in the foods* of various types which when enter into the body bring about nourishment. The nutrients are thus, materials and nutrition refers to the state of the body.

1.3.2. Types of Nutrients

Some specific chemical substances have been designated by the health scientists as nutrients, *i.e., components of nutrients.* These are six in number and are enlisted as follows:

1. Carbohydrates.
2. Proteins.
3. Fats (= Lipids).
4. Vitamins.
5. Minerals.
6. Water.

It should be noted that all the above components of nutrition are *chemical compounds.* They are present in various types of foods, whether as solid or as liquid. Hence, when we take solid or liquid foods, the nutritive components go inside our body.

In the body, the various nutritive components, *i.e.,* the constituents in chemical form are derived by the organs, in other words, the *alimentary canal* and thereby, the body gets the nutrients that are necessary.

As regard water, a layman may not consider it to be a chemical but it is very much a chemical compound and is chemically represented with the formula, which is H_2O. That is, in highly pure form (distilled water), it is composed of two molecules of hydrogen in chemical combination with one molecule of oxygen. Water which we drink, *i.e.*, potable water does not however, contain only hydrogen and oxygen but some other elements also in dissolved state.

All the nutritive components which come to the body through foods and drinks are in organic form except water itself, which comes in inorganic form. But water may enter into the body in organic form if a drink that contains organic compound(s) is taken.

It has been stated above that nutrients come to the body through the foods which are the *media of entry*. In this connection, it should be kept in mind that any single food does not provide all the nutrients and in right proportion. Making a combination of food, known as *mixed diet* becomes actually necessary to secure the different nutrients into the body in the desired proportion. The ancient Indian people knew this self-evident truth, *i.e.,* axiom without studying any book on nutrition.

Lastly, a fact is brought to notice, that is, nowadays a section of health experts are making attempts to thrust upon that a kind of food known as *dietary fibres* should be included under the group of nutrients. Their argument is, when the dietary fibres are of so great value on health, why not they should be regarded as nutrients. But the proposition has not been accepted by the majority of the nutrition scientists accross the world. It may be mentioned that another section of health professionals has suggested that some foods which have medicinal value should be considered as nutrients. Such a proposal has also been turned down by the nutritional scientists. As a matter of fact, unless a food item comes under one of the groups of nutrients as stated above, it should not be called a nutrient, regardless of what food value or merit it has, although there is vagueness of this argument.

1.3.3. Groups

The six components of nutrients that have been stated above are usually made into two groups. These are referred to as *macronutrients* and *micronutrients*. Those nutrients which are required in the body to a *considerably large amount* besides that, they have the ability *to release energy* on catabolism are termed by the nutritional scientists as macronutrients (macro = large). In that sense, the three nutrients, *viz.*, *carbohydrate*, *protein* and *fat* could justifiably be regarded as macronutrients as relatively high amount of them become necessary in the body and also that each of them may be catabolyzed (= enzymatically hydrolyzed) to liberate energy.

Some nutritional scientists consider *water* also as a macronutrient as it is required in much higher amount. However, *water does not produce any energy* like that of carbohydrate, protein and fat. Leaving aside these nutrients, the rest of the nutrients, *viz.*, vitamins and minerals are termed as micronutrients.

A group of nutritionists has classified nutrients as *essential* and *non-essential nutrients*. By essential nutrients, those nutrients are meant which are not produced in the body and accordingly, they have to be procured from external source, *i.e.*, food. The non-essential nutrients have been referred to those that are produced in the body. These have been elaborated in the subsequent Chapters of the title.

1.3.4. Benefits and Requirements

Nutrition of the body depends on what nutrients have gone inside the body and whether and how far, they are absorbed and utilized in the body. As a matter of fact, each of the six nutrients has to undertake specific functions in the body and when all the different nutrients are available in the body in optimum amount, proper growth takes place. The specific role of each nutrient, requirement and the amount recommended in accordance with age, gender, activity, state of the body and other conditions have been discussed in the subsequent Chapters.

In this connection, it is also brought to notice that some of the nutrients have been known to have *therapeutic properties* as well apart from their nutritive role. Thus, they act both as nutrients and as pharmaceuticals. Hence, a new term (= portmanteau word), *viz.*, *nutraceutical* has been introduced to such kind of nutrients by making a combination of the words, nutrients and pharmaceuticals. It may be mentioned that many of the nutraceuticals are widely prescribed to defend against or cure many an ailment. The advantage of the nutraceutical agent lies in that it can be taken through food.

1.3.5. Mode of Provision

Conventionally, the human body gets the various nutrients through the oral route, *i.e.*, food. However, for a sick person, it becomes necessary to deliver nutrients directly into parts of the body. Direct delivery of nutrients is done in one of the follwing three ways:

(i) *Enteral nutrition:* In this system, nutrients are directly delivered into the stomach, duodenum or jejunum but *in liquid form only* and through suitable tubes (= naso-gastric tubes) for incubated patients.

(ii) *Parental nutrition:* In this system, solution made with extremely fine salts of the required nutrients is injected into the blood stream intravenously.

(iii) *Total parental nutrition (TPN):* In case of gastro-intestinal dysfunction in a person, the entire nutritional requirement of the parents is administered in ultra-fine solution form through veins.

It should be remembered that the above methods of provision of nutrients directly into the body of the patient, which is known as *critical care nutrition* should be done with extreme precautions. Amount of nutrients which are to be delivered by these methods should be determined with great care as oversupply of nutrients in particular, may be fatal. Blood test of the patient to determine the nutrient status is advisable before the nutrients are delivered. Intravenous pushing of glucose or calcium solution or emulsion without heedful consideration to all possible aspects of the patient may bring about serious hazard.

Provision of nutrients is also done through medicines which are known as *nutritional supplements*. This is advised to a person only under special circumstances, *e.g.*, during growth of infants and children, illness, intake of poor food, excessive physical or mental activity, convalescence after illness, pregnancy, lactation, trauma, blood loss, old age and other conditions. Indiscriminate use of nutritional supplements and without advice by health providers or dietitians may cause great harm to the body. Such unnecessary intake of nutritional supplements which is done by some people without knowledge or being guided by incorrect advertisements may cause over-nutrition and harm to their precious health, besides emptying their pockets.

1.4. Malnutrition

1.4.1. Definition and Concept

Many people have the concept that malnutrition refers to *inadequate, i.e., insufficient* nutrition. That is, according to them, a person suffers from malnutrition means that he or she is not getting adequate nutrition what is necessary. In other words, he or she is undernourished.

The above notion is partly true but not wholly correct. This is because, in true sense, malnutrition is a *two-fold* nutritional problem. This points out that when a person suffers from inadequate nutrition, he or she could be said to be suffering from *under-nutrition*. On the other hand, if he or she gets too much nutrition, in other words, if he or she is bulked with so high nutrition which is not required to him or her, *i.e.*, get *over-nutrition (overnourished)*, then also the term, malnutrition will hold good.

In fact, malnutrition is defined as a *disorder of nutrition*. It is known to result under the following conditions. (1). If the nutrition is *not adequate* to whom it is provided. (2). If the nutrition provided is excess, *i.e., much more than what is required* by an individual. (3). If the foods or drinks containing the nutritional components are not *completely digested* in the body. (4). Even if digested but the nutrients that are present in the foods or drinks are *not adequately absorbed* in the body. (5). Even if absorbed, the nutrients are *not transported to* concerned organs where they are actually needed. (6). If the person is *not physically* or *mentally benefitted* on getting the nutrients.

The six components of nutrients as stated above under the sub-heading, 1.3.2 are essentially required to the body for growth, to get energy and to maintain proper health. Further, it has been described in the subsequent Chapters that except water (which is a nutrient), other nutritional components, *viz.*, carbohydrates, proteins, fats, vitamins and minerals are only broad groups and within each of these groups, there are many sub-groups, which are the actual nutrients. Some of these nutrients coming under some of the broad groups are required to the body in very less amount and some are necessary in extremely small amount. However, such a nutrient which even if required in extremely small amount, *i.e., in traces only has also its essentiality* for growth and maintaining proper health of the body. Some vitamins and minerals are examples of such type and if any such vitamin or mineral that is required in minute amount, i.e., in *physiologically active concentration* is not received by the body, there will be a set-back of health, which will result in malfunctioning of the body systems, effectuating improper health. Evidently, such a situation comes under one of the two sides of malnutrition. It may be pointed out in this connection that when deficiency of some of the nutrients goes to the extreme level for some period, the body should be regarded to be in a state of *starvation*.

The other side reflects that not only deficiency of any particular nutrient does give rise to ill-health but if any of the nutrient goes inside the body to such an amount which is more than what is actually necessary, the state of the body should then also be regarded to be in malnutrition. This over-supply of nutrients to the body is termed as *hyperalimentation*. Apparently, this would appear to be a fallacy particularly to the glutton (!) but the physiological condition of his or her body knows what harm he or she does to the precious health just by over-eating. Mahatma Gandhi had said, "Over-eating kills more people than those who eat less".

1.4.2. Symptoms

Malnutrition in an individual exhibits symptoms especially when the situation persists for some period. The undesirable deviations from the normal condition are found to occur in outer appearance of the body, in the internal organs and even affect mental health in a person. It is of interest to note that many children do not apparently show any sign of malnutrition but on critical examination, they may be found to have been suffering from malnutrition. An example may be cited as impaired mental growth and attitude of some children. Rapid progress of brain development which may be upto an extent of 90 per cent takes place in children from 3 to 6 years of their age. If the nutrition is not adequate at this age, a child may be seen to carry on physical activities well or to a moderate level but he or she may develop mental lacunae. These undesirable signs may be many types such as phobias, too much shyness, abnormal symbolism, frequent urination, cognitive dysfunction, poor memory, apathy towards arithmatical works and many other types of mental illness. If may be added in this connection that *glucose is a suitable fuel to the neurons and brain* but this aspect is perhaps not given much attention by many health experts.

Individual nutrients are plenty in number within the six broad components of nutrients. Each component has specific roles in the physiology of the body and even a slight alteration from the exact requirement of any nutrient will cause malfunctioning of one or more body systems, eventuating production of undesirable symptoms. Deficiency or excess of a given nutrient shows symptoms of specific types and these have been stated in the subsequent Chapters. However, some of the common symptoms due to inadequate nutrition may be stated as stunted growth, underweight and wasting of children that are not proportional to their age, anaemia, scurvy, blindness, skin abnormalities *etc.* in adults and giving birth to babies prematurely may be observed in women due to undernourishment.

Most common symptoms of super-optimum intake of some nutrients notably, fat, carbohydrates (= monosaccharides and oligosaccharides) and protein are obesity, metabolic syndrome, type-2 diabetes, high level of cholesterol and triglyceride in blood, respiratory and cardiovascular disorders *etc.* Some types of cancerous growths have also been stated to be an effect of over-nourishment. Excess intake of protein may produce symptoms of urological disorders.

1.4.3. Causes

The causes of malnutrition may be either *external* or *internal* or both. External factor is, intake of faulty, *i.e.*, improper food, while the internal cause relates to lack of utilization of the nutrients that actually go inside the body.

Intake of nutritionally inadequate diet occurs for a number of reasons and the more common ones may be stated to be as follows: Poor economic condition is a major factor among some people for which they cannot afford to obtain foods of high nutritive value. Although meat, fish and egg are rich sources of protein of high quality, many people cannot consume them regularly and in sufficient amount for their high cost as compared to vegetarian foods. In India again, considerable part of the population refrains from taking meat, fish and egg only on religious ground. However, the milk, which is though a non-vegetarian food is taken by the people belonging to the Hindu religion and hence, they are regarded as lacto-vegetarians (= milk-vegetarians), rather than completely vegetarians in food habits. There is a report which states that the people "belonging to the Hindu or Muslim backgrounds in India tend to be more malnourished than those from Christian, Sikh or Jain backgrounds".

Geographical region is known to be another factor for under-nourishment of people in India to some extent. Thus, a considerable section of people in Madhya Pradesh, Jharkhand and Bihar is observed to be more malnourished than people of other regions. Percentage of population afflicted with anaemia is known to be more than 70 in the States like Chhattisgarh, Madhya Pradesh, Andhra Pradesh, Jharkhand, Bihar *etc.* and less than 50 in Goa, Kerala, Manipur *etc.* As compared to under-nourishment, the primary cause of over-nutrition in people of India is high socio-economic status. Lack of knowledge on nutrition is also a great factor as many people in the country take too much meat, fish and egg for their wrong belief that consumption of animal source protein will keep them and their children more healthy and intelligent.

Malnutrition in a person may take place due to internal factor as well, *i.e.*, due to physiological condition of the body. It has been stated earlier under the heading, 1.1 that nutrition in the body systems continues in several stages or steps, which are, digestion, absorption, transport and their assimilation. If there is malfunctioning in any of these physiological processes, nutrients that have been taken inside the body through food will not be utilized and accordingly, bring about malnutrition. Malfunctioning of physiological process may occur due to ill health, pathogenic infection, physiological disorder or inherent abnormalities. Hepatic and pancreatic disorders are very common cause that stand in the way of ultilization of nutrients in the body. Children attacked by intestinal worms are often found to suffer from malnutrition despite intake of nutritive foods.

A section of nutritional scientists is of the opinion that even psychological factors play part in malnutrition of many people. Ackerson, L. K. (2008-05-15) [in their article, "Domestic Violence and Chronic Malnutrition among Women and Children in India, "published in the American Journal of Epidemiology, Vol. 167 (No. 10), Page No. 1186 – 1196] stated, "A strong connection has been found with malnutrition and *domestic violence*, in particular with high level of anaemia and undernutrition. "An article [published in the Journal, Social Science and Medicine, Volume, 72, No. 9 (2011) titled, "Impacts of Domestic Violence on Child Growth and Nutrition : A Conceptual Review of the Pathways of Influence, went on to quote, " Domestic Violence comes in the form of psychological abuse as a control mechanism towards behaviour within families. This control exerts effect on a woman's autonomy in making decisions in respect of providing type and amount of food and leads to adverse effect on herself and other members of her family.

Other nutritional scientists have maintained through published articles that mental tension produces oxidative stress in the women particularly in India and due to *oxidative stress, free-radicals* are produced. *These free-redicals attack and damage the healthy red blood cells*, for which red blood cells are reduced in number and eventually, they become anaemic, besides being victims of many unhealthy conditions.

1.4.4. Assessment

Whether or not a person is suffering from chronic form of malnutrition is an issue of health that demands keen attention. However, how to assess the nutritional status of a person with exactness is a difficult task also but is not impossible. Assessment may be done in a number of ways and those which are commonly employed may be stated as follows:

(i) General Appearance

It is a tendency among many people to call a person healthy or sick just by observing him or her outwardly. A person who is not very lean or very fat is considered to be in good health and accordingly, it comes to the mind of the observer that the person is regularly taking nutritive food in adequacy. In India, a child who is fat is called to be in good health especially if he is a boy rather than a girl child and it is assumed that he takes adequate nutritive food every day. Judging a person to be healthy or not and whether he or she is nutritionally all right just on observing external appearance of him or her is although very common in India can hardly be regarded to be a correct assessment and *this is often misleading*. When a fat (roly-poly) boy is called by us to be in good health, we forget that the extra fat in his body is actually a sign of bad health and in most cases, this unwanted fat is a consequence of nutritionally faulty diet, although there may be other causes (*e.g.*, thyroidism, less physical activity *etc.*) also for fatness. Even many health providers consider outward appearance of a person as the only criterion to judge his or her health and nutritional adequacy in his daily diet.

(ii) Weight

Very commonly, weight is considered to be an index to judge the general health of a person. A health expert also takes the weight of the patient after feeling the pulse rate, observing the tongue and stethoscopic examination of the chest and back. The weight is considered to be an important criterion of good or bad health, emaciation and enfeebling. In case of underweight of an individual in absence of any serious morbidness, whether malnutrition is the causative factor comes to the mind of the physician or the dietitian. However, although weight of a person is much a dependable index to judge whether he or she is suffering from malnutrition or not, it could hardly be said to be much correlated with nutrition.

(iii) Body Mass Index

Instead of weight alone, both weight and height taken together is a more effective indicator to assess general health and nutritive status of a person.

Making a combination of height and weight (= mass) collectively, the idea of *Body Mass Index* (BMI), the other name of which is *Quetelet Index* has come into being. The Body Mass Index of a person can be determined easily as stated below:

The person is allowed to stand in an absolutely straight position along a wall and preferably at the corner of the wall. He or she should stand bare-footed, *i.e.*, should not wear shoes, slippers and socks. At this position, his or her height should be measured accurately. Height should be measured in metric units only (= metre, decimetre, centimetre) and not in feet and inches. The height which is measured is made squared. That is, if the height is measured as say, 1.65 metres, it should be made squared (1.65 m × 1.65 m) so as to get the product as 2.72 square metres.

Then the weight of the person should be measured precisely and in metric units only (= kilogram, hectogram, dekagram). To get the exact weight of the body, evidently the person should not wear heavy weight garments.

On getting the weight and square metres, the BMI can be worked out from the following formula:

$$\text{BMI} = \frac{\text{Weight (in metric unit)}}{\text{Height (in square metres)}}$$

That is, if the actual height is 1.68 metres and if it is made squared, the product will be 2.82 square metres. If the weight in 73.00 kg, the BMI will be 73.00 divided by 2.82, *i.e.*, 25.89.

On determining the BMI, inference should be made in consideration to recommended or much acceptable standards. A standard reference is as follows: (i) BMI from 15 to 16 = severely underweight, (ii) 16 – 18.5 = underweight,

(iii) 18.5 – 25 = normal, (iv) 25 – 30 = overweight, (v) 30 – 35 = moderately obese, (vi) 35 – 40 = severely obese, (vii) 40 – 50 = very severely obese, (viii) 45 – 50 = morbidly obese, (ix) 50 – 60 = super-obese and (x) above 60 = hyper-obese. (Ideally, the BMI should be around 23 for the Indian people).

Examining the overall health of a person on determining the Body Mass Index has gained much importance now. The physicians, nutrionists and the dietitians are adopting this test as one of the first line of judging the general health or diagnosing a chronic illness in particular of a person. The test can be done quickly and is a reliable index to determine health of a person for many a case. Nevertheless, there are limitations in this method and the major drawbacks may be stated as follows:

Body Mass Index largely gives an index of body fat although gives some indication of the protein and fat mass of the body. The equation gives an approximation of body fat only and does not determine the fat content that exists in different organs. This is of significance as deposition of fat in some organs may be of much concern. Can it be said that *women deposit fat under the skin while men deposit within the veins and arteries for which they are more prone to atheroselerosis*?

BMI does not consider waist circumference which is an important factor for the Indian people who are ridiculed as having *apple-shaped bodies* (or 60 inches persons). (According to Prof. Gita Mathai of the Christian Medical College in Vellore in Tamil Nadu, in women, the waist circumference should be a maximum of 35 inches, *i.e.*, 88.9 cm and in men, the maximum circumference should be 40 inches, *i.e.*, 101.6 cm). A certain age-range in an important factor for determining BMI in a better way and is not very correct for persons of all ages. BMI cannot give a correct index of the nutritional status of a person and only gives a general indication only and is not of much value.

(iv) Anthropometric Tests

This is a test which involves physical measurements of body parts. The test is of value to determine general health of an ailing, convalescing or a healthy person but does not appear to be much useful to determine nutritional status of a person. Anthropometry (Anthropo = human beings) includes craniometry (= study of skull and measurement of bones), osteometry (= study of measurement of skeletal parts), skin fold evaluation to determine deposition of subcutaneous fat as well as weight and height. (The tests are done by anthropometric specialists). The tests are not considered very reliable.

(v) Symptomatic Evaluation

It is very common among the physicians to judge the general health as well as nutritional status of a person on observing symptoms as the first step of treatment. Expert nutritionists also attach much importance to observe symptoms. It may however, be stated in this context that determining health condition and nutritional status of a person may although be done with fairly high degree of accuracy for some ailments and in case of deficiency or excess of some nutrients, precise assessment is difficult to be made in many cases. This is because, symptoms may be alike for more than one health abnormality and that, many symptoms or syndromes may appear for the same abnormality. It may be said that anaemia may be judged by observing the paleness of the eyelids (palpebra, canthi), Grave's disease on observing eyes but correct assessment on observing only on symptoms is far from diagnostic approach and may be highly misleading but nevertheless, cannot be ruled out. For example, appearance of yellow patches on skin surface is a common symptom of jaundice but such a condition is also observed if a person takes high amount of carotene (vitamin A precursor) for a considerable period of time, which is known as *carotenemia* or *xanthosis*. An expert health scientist knows that unlike in jaundice, the eyes do not look yellow in case of carotenemia which is only a malnutrition disorder, caused by excess intake of carotene. In the subsequent pages which included the study of nutrients, the symptoms of deficiency and excess of different nutrients (macro- and micro-) in other words, malnutrition symptoms have been discussed and hence, avoided in this chapter.

(vi) Blood Test

Nutritional status of a person at any given time may be known by qualitative and quantitative estimations of the various nutrients that are existent in a sample of blood carefully taken out from his or her body. The method gives a correct picture and is highly reliable. However, there are limitations in this method and some may be stated as follows:

The method involves very high cost and is hardly affordable by a person every now and then and hence, is advised by the physicians or the nutritionists when such a necessity is absolutely felt. That is, when diagnosis becomes difficult on observing the externally perceived symptoms only or when the medicines administered do not respond well, the patient is advised to have blood test to determine the status of at least some of the nutrients. Cost of the test goes still higher when some of the amino acids, vitamins and micronutrients which are present in extremely small amount are needed to be estimated from the blood plasma. It may be stated that sophisticated facilities for conducting all the tests are in fact, not available everywhere. Correct assessment on making comparison with standard reference tables sometimes becomes difficult as the reference values that are available have been obtained on making analyses which have been performed in altered physiological condition of an individual. It may further be commented upon that free floating of nutrients in the plasma fluid does not always mean how much of them would actually be absorbed, assimilated and utilized in the body. Despite other practical difficulties, occasional examination of blood, sputum, urine and stool is a practice which is done among defence personnels to rule out pathogenic infection and malnutrition.

Lastly, it may be opined that instead of one test, a number of tests become necessary to have a clearer picture in regard to the nutritional status of a person and to judge whether he or she is suffering from malnutrition of any kind.

(vii) Extent

Malnutrition is a health problem that exists all over the world. According to the World Health Organization's report (2016), 462 million adult persons around the world are underweight and among the children below the age of 5 years, 159 millions are stunted in growth while 50 millions are wasted. On the other hand, 1.9 billion adults are overweight and among the children under the age of 5 years, 41 millions are overweight, *i.e.*, obsessed. Thus, malnutrition is a *dual burden in that, both under-nourishment and over-nourishment are problems.* Number of persons suffering from malnutrition widely differ in countries.

Chapter 2

Carbohydrate

In the Chapter 1, the six components of nutrients have been stated as carbohydrates, proteins, fats, vitamins, minerals and water. Among these nutrients, *carbohydrates may be ranked as the first*. This nutrient which makes up large part of the body is the theme of discussion in the present Chapter.

It may be stated at the beginning that the human beings have come to know the taste of *sweetness* in ancient time when they ate ripe fruits of date, fig and grape and put honey on their tongue. Later, they learned the technique of securing sap on tapping that from the trunk of date-palm. It came to their knowledge that the Nature's gift, the sweet materials not only provided them a gustative appeal to their tongue but also could provide quick energy for doing physical work. Thus, the importance of sugar which is a carbohydrate was known to them.

With rolling of ages, the ancient people started cultivation of crop plants and they came to understand the benefits of the carbohydrate, *i.e., starch* which was derived from the cereals and the underground-grown tubers. This was used by them as food to fulfil their stomach, for use as fodder for the reared animals and to use as a sticking material, *i.e.*, glue in building their hutments. Thus, without having the knowledge of chemistry, the ancient people had discovered carbohydrates and realized the various benefits.

In the Sanskrit literature in India, carbohydrate had been referred to as *sharkara* and the ancient people of India had a high regard to that. Even now, food-stuffs rich in starch are valued and offered as oblation to the gods and goddesses in Hindu religion. The word, saccharide which is used in chemistry has its root in Greek, *sakharon*, meaning of which is sugar.

At the present time, staple food of all people across the world is cereals and millets, which provide high starch in the body. The global production of cereals and millets like rice, wheat, maize, sorghum and other small millets is now estimated to about 105, 95, 250, 6 and 2 million tonnes. These grains are *rich in starch* and apart from consumption as primary source of food, a large number of processed products are also made from them.

Carbohydrates serve as essential *base materials* in the synthesis of many organic compounds in the body that are of vital necessity. A group of carbohydrates known as polysaccharides (= glycans) may be mentioned which are of great benefit to maintain health and keep at bay many ailments. These compounds termed as *dietary fibres*, which are secured from plant source foods have been discussed later.

Many forms of carbohydrates, *e.g.*, glucose, sorbitol, pectin, starch, cellulose *etc.* or the foods that are rich sources of these carbohydrates are often recommended by the physicians, nutrition advisers and dietitians to defend against many maladies. These are advised to prevent or cure many types of pathogenic infections and insalubrites. For these qualities, they are termed as *nutraceutical* agents. Of late, carbohydrates of some types are known to be efficacious in some types of psychic illness, depression, inquietude and in various types of dipsomania. Some of the pertinent health benefits of carbohydrates have been brought to notice later under the heading, 2.8.

Apart from those stated above, it may be mentioned that many of the carbohydrate compounds serve as ingredients, either singly or in suitable combinations with other substances in a number of industries, such as in food, confectionery, bakery, ice-cream, distillery, pulp, dyeing, cosmetic and pharmaceutics.

Our surroundings are largely filled with the carbohydrates and there is no denying the fact that life without these chemical substances cannot be thought of. Looking into the various forms of chemical compounds that exist in the body of living beings, it would be apparent that they come under four major groups and in chemical terms, these are recognized as *protein*, *carbohydrate*, *fat* and *nucleotide*. Among these, however, carbohydrate is the *most abundant material*. This chemical compound is the primary organic substance on Earth and is the invaluable material to provide power to rotate the wheel of human civilization.

2.1. Definition

The word, carbohydrate is derived from the words, which are *carbon* and *hydrate*. This apparently denotes that a molecule of carbohydrate is chemically such a compound that it contains carbon atom(s) and this exists in chemical combination with hydrate(s) radical. That is, the hydrate is separate but it remains in a *chemically combined form* with that of hydrogen and oxygen.

Such a term was proposed by the chemists many years ago from the fact that glucose, which was the first carbohydrate that had been obtained in pure form was found to have the molecular formula as $C_6H_{12}O_6$. It was conceived that glucose could be a hydrate of carbon and hence, its actual formula would be $C_6(H_2O)_6$.

This concept was held for a long period but later, the scientists came to understand that the proposition was not correct. They put forward the fact that although it is true that carbohydrate molecules are basically composed of three chemical elements, *i.e.*, carbon, hydrogen and oxygen, their configuration is in fact, not arranged in such a manner that they chemically behave as hydrates of carbon. However, despite the fact that the term, carbohydrate does not really seem to be appropriate, it is still retained in the science of chemistry.

A carbohydrate molecule has three essential elements, which are carbon, hydrogen and oxygen and they make plentiful combinations. Hence, a huge varieties of carbohydrates are found to exist on Earth.

In the next heading, discussion has been made on the general structure of the carbohydrates and proportion of carbon, hydrogen and oxygen that are present in the molecules of different types of carbohydrates. These apart, the scientific reasons of *why carbohydrates should not be regarded as hydrates of carbon* have been brought to light from chemical point of view.

2.2 Structure

In a carbohydrate molecule, the three elements, *viz.*, carbon, hydrogen and oxygen are most commonly arranged in the ratio of 1 : 2 : 1. That is, along with one atom of carbon, two atoms of hydrogen and one atom of oxygen remain in chemical combination. Thus, if carbohydrate compounds are expressed with the formula, $Cn(H_2O)_n$, where n represents number, it may appear that in a carbohydrate molecule, hydrogen and oxygen atoms exist in the proportion of 2 : 1. It may be pointed out in this connection that in a water molecule also, *i.e.*, H_2O, proportion of hydrogen and oxygen remains same, which is 2 : 1. Accordingly, it seems apparent that a carbohydrate molecule is such a one that each carbon atom of it is chemically combined with at least one molecule of water. Thus, it seems that a molecule of carbohydrate is actually a *hydrate of carbon*.

However, such a postulation that a carbohydrate molecule is a hydrate of carbon which was conceived by the chemists of the yesteryears has not gained support by the chemists of the subsequent years. The following four reasons have been brought forward for disagreement of this concept.

Firstly, although in a large number of carbohydrates belonging to different groups, proportion of hydrogen and oxygen atoms in their molecules remain as 2 : 1 but such an arrangement of these two atoms is *not a must for carbohydrates of all types* and is not a universal phenomenon. For example, if we take the example of *rhamnose* which is a carbohydrate, we see that in its molecule, hydrogen and oxygen atoms are not present in the ratio of 2 : 1 like that of a water molecule. The chemical formula of rhamnose is $C_6H_{12}O_5$, in which, hydrogen and oxygen atoms *do not exist* in the ratio of 2 : 1.

Secondly, in a carbohydrate molecule, even if the hydrogen and oxygen atoms exist in the ratio of 2 : 1 but these are *not held in chemically combined form with carbon*. That is, a carbohydrate molecule should not be chemically considered to be a hydrate of carbon. Had it been so, when a carbohydrate molecule is heated, water molecule(s) should have been evaporated, leaving behind only the carbon molecule(s). But that does not happen.

Thirdly, there are some organic compounds whose chemical formulae are same as carbohydrates but they are not actually carbohydrates. Some examples of such compounds may be cited as formaldehyde (CH_2O), acetic acid ($C_2H_4O_2$) *etc*. These compounds have same chemical (= empirical) formulae as that of carbohydrates but they do not come under any group of carbohydrates.

Fourthly, there are some carbohydrates, whose molecules not only contain carbon, hydrogen and oxygen but other elements like nitrogen and sulphur are also present in them.

To define what carbohydrate is from chemical standpoint, we have in fact, to consider chemistry of *aldehyde* and *ketone*. Aldehyde radicals (= IUPAC name : alkanals) (–CHO), *i.e.*, groups are those which have double-bonded carbon and oxygen atoms with one hydrogen atom while ketone radicals (= IUPAC name : alkanones) are double-bonded carbon and oxygen atoms with two additional carbon atoms (>C = O). In true sense, all carbohydrates are aldehyde or ketone *derivatives* of polyhydric alcohols of the aliphatic series and the products of their condensation. Therefore, in appropriate terminology, *carbohydrates are aldehydic or ketonic derivatives of polyhydric alcohols, i.e.*, which have more than one hydroxyl (= OH) groups.

2.3. Functions

The most important function of carbohydrate in the science of nutrition is to *provide energy in the body*. Many physiologists of the present day opine that although protein is considered as the first in nutrition (= the word, protein is actually derived from the Greek word, *protos* which means first) but in fact, glucose should deserve this rank as this is an energy-giver and it is energy that capacitates growth and absence of it brings life to a standstill. Energy is liberated by the monosaccharide forms of carbohydrates by fuelling ATP synthesis and also in making chemical reduction of NADPH. Glucose molecules liberate energy without giving rise to any metabolic waste, *i.e.*, it breaks down to produce carbon dioxide gas and water only. The dipeptide, tripeptide and polypeptide forms of carbohydrates also produce energy but they do so *on breaking down to glucose molecules only* and not directly. It has been stated later that 1 gram of monosaccharide provides 3.74 Kcal, 1 gram of disaccharide 3.95 Kcal and 1 gram starch provides 4.18 Kcal of energy in the body.

Sparing protein to liberate energy is another very important attribute of carbohydrate (and also fat). Protein is a highly valuable substance and it performs many important functions in the body. But under certain conditions, protein molecules are broken down to liberate amino acids and on doing so, they give rise to energy. As a matter of fact, this is a *misuse of protein*. For providing energy in the body, carbohydrates are ideal substances. (Fat may also be regarded for the purpose). One gram of protein gives rise to 4.2 Kcals (= 17.57 kilojoules) of energy and 1 gram of carbohydrate also provides the same energy. (One gram of fat provides 9 Kcals, *i.e.*, 37.8 kilojoules). Therefore, for the purpose of getting energy in the body, carbohydrates should be regarded as *ideal materials*. But this would happen when the carbohydrate reserve in the body is adequate. Thus, carbohydrates could be said to *spare the use of protein* for the purpose of getting energy. This has been discussed in Chapter 3 (Protein) under the heading, 3.12 as *protein sparers.*

Carbohydrates not only provide instant energy in the body but they are also capable to make provision of energy later when there is necessity. In fact, they do so by the act of *storage*. A polysaccharide form of carbohydrate, known as *glycogen* is the material which does this function. This compound is a highly rich source of glucose molecules. Glycogen is adequately stored in the *liver* and *muscles*. When energy is required in the body by glucose, (which is the source of energy) is not adequate, glycogen is enzymatically broken down (= catabolized) to give rise to glucose molecules, *i.e.*, the simplest sugars and thus, provide energy. In this connection, it may be brought to notice that in the life of plants, *such a rich source of energy* that does the same function is *starch*. This polysaccharide form of carbohydrate could therefore, be regarded to be a counterpart of glycogen in the body of a plant.

There are many other functions of carbohyrates which have been mentioned later under the heading, 2.8. Nevertheless, a few are stated below:

Carbohydrates are chemically linked to many proteins and lipids. They have detoxifying properties. The two pentose sugars, ribose and deoxyribose are components of nucleic acids. Glucose maintains integrity of nervous tissue and is the proper fuel for the functioning of brain. Cellulose provides structural support to plant cells (protoplasm) and chitin to exoskeleton (= outer hard covering of the body) of insects and other arthropod animals. When carbohydrates accumulate in excess, they form acids and fats.

2.4. Chemical Properties

Carbohydrates posses a large number of chemical properties. A few characteristic features of the *monosaccharides* which are the most simple and primary forms of carbohydrates are stated in the following. Some of the chemical properties of important carbohydrates have been brought to light under the next heading (2.5.2).

(i) *Sweet taste* is by far, an attribute of the carbohydrates of the monosaccharide group. Sweetness may be low or high depending on the type of monosaccharides. Considering sucrose (= common sugar) to have a *score* of 100 marks, the relative sweetness (not in any unit) of some monosaccharide and disaccharide sugars as well as some synthetic sweeteners have been stated in Table 1.

Table 1. Relative Sweetness of some Natural Sugars and Synthetic Sweeteners (Considering sucrose to score 100)

Compound	*Type (Group)*	*Relative Sweetness (Not in any unit)*
Lactose	Disaccharide	16
Raffinose	Trisaccharide	22
Galactose	Hexose	32
Maltose	Disaccharide	32
Xylose	Pentose	40
Glucose	Hexose	74
Sucrose	Disaccharide	**100**
Invert sugar	Glucose + Fructose (1 : 1)	130
Fructose	Hexose	173
Sodium cyclamate	(Synthetic)	3000
Saccharin	(Synthetic)	45,000

(A trisaccharide, *viz.*, melezitose, syn. melicitose is known which is fairly sweet in taste. This is produced by plant sap-eating insects by enzymatic action. This is a part of the honeydew which is an attractant for ants and a food for the bees).

(ii) Monosaccharides have *no colour* or *odour* of their own.

(iii) These are derived from *polyhydric alcohols, i.e.,* alcohols having many hydroxyl groups. Because of many such groups, these are easily soluble in water but insoluble in organic solvents.

(iv) Aldehyde or ketone group of aldohexoses and ketohexoses can be *chemically reduced* to form the respective polyhydric alcohols. (Reduction of glucose leads to sorbitol and reduction of fructose produces sorbitol and mannitol).

(v) *Isomerism* and *optical activity* are important properties of monosaccharides. This takes place due to asymmetrical carbon atoms that are present in them. When a carbon atom has four different types of atoms or groups of atoms in its four valency bonds, it is called *asymmetric*. Such an asymmetric atom becomes optically active and the compounds of such atoms exist in many forms which are called *isomers*. In isomers, although molecular formulae remain same, their spatial configuration is different. The isomers exhibit optical isomerism. This phenomenon indicates that when *a beam of polarized light, i.e.,* a beam of light that passes through a single plane is made to pass through a solution of a chemical compound, it *rotates the light either to the left or to the right direction*. Such a compound whose solution has this property is termed as *optically active*. If the light rotates to the right, it is called *dextro-rotatory* and is designated by

(i) *Biose* – which contains 2 carbon atoms, *e.g.*, glycolic aldehyde.

(ii) *Triose* – which contains 3 carbon atoms, *e.g.*, dihydroxyacetone.

(iii) *Tetrose* – which contains 4 carbon atoms, *e.g.*, erythrose, erythrulose.

(iv) *Pentose* – which contains 5 carbon atoms, *e.g.*, ribose, arabinose.

(v) *Hexose* – which contains 6 carbon atoms, *e.g.*, glucose, fructose.

(vi) *Heptose* – which contains 7 carbon atoms, *e.g.*, sedoheptulose, mannoheptulose.

The series continues in this manner.

(b) According to *aldehyde or ketone groups* that are present. That is, when a monosaccharide molecule contains an aldehyde group (– CHO), it is called aldehydic sugar or *aldose*. Similarly, when the monosaccharide molecule contains a ketone group (C = O), it is called ketonic sugar or *ketose*. Glucose is an example of aldose and fructose is an example of ketose.

(II) Oligosaccharides

The term, *oligo* in Greek means few. Thus, when a carbohydrate molecule contains more than one unit of sugar (= saccharide), it is called oligosaccharide. That ... in an oligosaccharide molecule, *more than one unit (molecule) of monosaccharides* ...ain present (= *joined together*). Two questions apparently come in this context. ...are:

how many monosaccharide units (molecules) should actually be joined ...gether to form an oligosaccharide unit (molecule) and

... the joining of the monosaccharide units take place to form an ...charide. This is discussed in the following:

...ard number of monosaccharide units that are necessary t... ...d together to form an oligosaccharide, correct answ... ...le. (Oligo means few and few does not state any ... ...ually, controversy exists as to how many non-... ...ecessary to be joined together. A lar... ...pinion that in an oligosaccharid... ...saccharide units should b... ...e joined together, it ... ...position is how... ...emists, th... ...gosa...

...charides.

...one mo...

+ (positive) sign and when rotates to the left, it is called *laevo-rotatory* and is designated by – (negative) sign.

(vi) On oxidation, carbohydrates form *sugar-acid*. Oxidation of glucose gives rise to sugar-acids of three types, *viz.*, gluconic acid, glucuronic acid and glucaric acid.

(vii) Hot and diluted mineral acids do not react with monosaccharides. Carbohydrates are *dehydrated by concentrated acids* and form furfural.

(viii) Monosaccharides are good *reducing agents in alkaline medium* and thereby, they are capable to reduce oxidizing ions, *e.g.*, those of silver, copper, bismuth, mercury *etc.*

(ix) Formation of *glycosides* is a feature of the sugars, *i.e.*, the sugar part (= glycone) of the carbohydrate is bound to be a non-sugar (= aglycone) part.

(x) Chemical esters, *i.e.*, organic salts (IUPAC names : alkanoates) are easily formed from monosaccharides. It may be stated that among different esters, *sugar-phosphate esters* are of great physiological importance. Esters and their derivatives are soluble in organic solvents and these are purified easily and crystallized.

(xi) Monosaccharides are crystalline carbohydrates and hence, the molecules have definite geometric shape, *i.e.*, they are bounded by plane surfaces.

(xii) Many of the monosaccharides are used as *substrates*, *i.e.*, food of micro-organisms of different species. The micro-organisms grow on them and respire *anaerobically*. As a result, organic acids, alcohols *etc.* are produced as end-products and carbon dioxide gas is given off. This gas comes out penetrating the monosaccharide substrate for which it becomes porous like a bread (= loaf). The process is known as *fermentation*.

(xiii) *Osazone* formation is a character of the monosaccharides. This means that when a monosaccharide is reacted with the chemical, *phenyl hydrazine* ($C_6H_5NHNH_2$, which is used to prepare indole), it forms a yellow coloured product, which is known as osazone. At first, phenyl hydrazine reacts with sugar to form *hydrazone* and then it is oxidised to form *glucosazone* (= osazone).

(ix) *Hexosamine* formation is another property of the monosaccharides. This takes place when the *hydroxyl* (OH) group of a hexose (C_6) sugar is replaced by an *amine* (NH_2) group. Thus, the hexose sugar on reaction with amine forms *hexose-amine* (= hexosamine).

(x) Monosaccharides exist as *straight-chain* compounds and also as of *ring-forms*. When in ring form, the ring may be of six members or of five members. In the six-membered ring, there are five carbon atoms in the ring along with one oxygen atom. In the five-membered ring, there are four carbon atoms in the ring with one oxygen atom. Haworth proposed that all sugars which are of six-membered rings are to be termed as *pyranoses*, because of their relation with *pyran* (= pyran or oxine is a six-membered, heterocyclic, non-aromatic ring, C_5H_6O), while all those sugars which are of five-membered rings are to be termed as *furanoses* because of their relation with *furan* (= furan is a heterocyclic, aromatic ring, C_4H_4O).

2.5. Classification

There are plentiful types of carbohydrates. Although they are all recognized as carbohydrates, they differ amongst themselves on many respects, such as chemical structure and configuration, nature, properties *etc.* Accordingly, their classification becomes necessary to distinguish them from one form to the other.

Classification of carbohydrates has been made in many ways. Long back, taste focused attention to classify them. Accordingly, those carbohydrates which when taken in mouth gave sweet taste were regarded as *sugars* and those which lacked sweetness were considered as *non-sugars*. This type of grouping is not prevalent now.

Another classification was done earlier in consideration to simple or complex nature of them. Thus, those carbohydrates, which are not possible to be converted (broken down) into still simpler carbohydrates by *hydrolysis* were recognized as *simple carbohydrates* or simple sugars, *e.g.,* glucose, fructose *etc.* On contrary, those carbohydrates which are possible to be broken down into two or more simple carbohydrates by hydrolysis had been termed as *complex carbohydrates*, *e.g.,* sucrose, starch, cellulose *etc.* Thus, in a complex carbohydrate, the two or more simple carbohydrates remain chemically joined (= linked) together. This classification is also not much followed now.

At the present time, organic chemists across the world prefer to clas carbohydrates into the following three broad groups. This traditional cla is largely accepted now.

(I) Monosaccharides,
(II) Oligosaccharides, and
(III) Polysaccharides.

Their characteristics and sub-groups are state

(I) Monosaccharides

The word, *mono* means one and the Greek word, *sakharan*, which means s one molecule of sugar and this cann still other carbohydrate.

Monosaccharides are grouped in tw

(a) According to *number of carbon at* grouped according to the number o molecule of monosaccharide. The sub-g

out). When two monosaccharide units are thus, joined, it is called a *disaccharide*. Joining of two monosaccharides to form a disaccharide molecule by glycosidic linkage has been shown in Figure 1. Like a disaccharide, when 3 monosaccharide units join together, the process (= glycosidic linkage) remains similar and the joined product is called a *trisaccharide*. The same process continues when more than 3 monosaccharide units are joined.

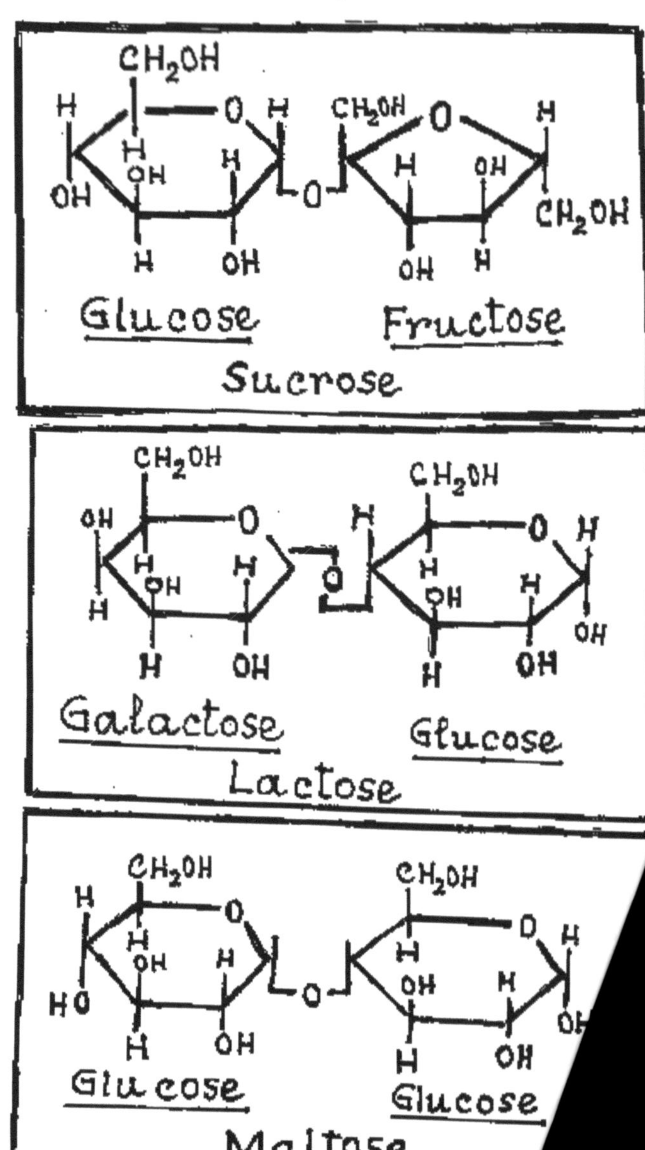

Figure 1. Formation of Three Types of Disa

It should be noted that instead of using the term, *joining* of monosacchrides, other terms, such as linkage, condensation, bonding, combination, attachment, unification *etc.* of monosaccharides may also be used. It should also be noted that as the individual monosaccharides join to form the oligosaccharide units, each individual monosaccharide is called a *unit* or a *monomer* and the product which is formed by joining the monomers is called a *polymer*. Thus, a polymer is a repeating unit of monomers.

The oligosaccharides are classified, *i.e.*, sub-grouped according to the *number of monosaccharide units* (= monomers) that are joined in it. Thus, they are sub-grouped as (i) *disaccharides* when only 2 monosaccharide units are joined together as in sucrose, lactose *etc.*, (ii) *trisaccharides* when 3 monosaccharide units are joined together as in raffinose, maltotriose *etc.*, (iii) *tetrasaccharides* when 4 monosaccharide units are joined together as in stachyose, sesamose *etc.* and so on.

It is pointed out in this connection that formerly, the disaccharides were brought under a *separate group* and not under oligosaccharides. But a large section of chemists is not in favour of that and according to them, it is more logical to place the disaccharides under the group of oligosaccharides. Nevertheless, a section of organic chemists still prefer to make a separate group for the *disaccharides.*

(III) Polysaccharides

A polysaccharide (poly means many) is a carbohydrate which is formed when more than 9 (or 10) monosaccharide units are joined together. There is no upper limit and as many as 107 or more monosaccharide units may be joined together to make one unit (= molecule) of a polysaccharide. Joining of the monosaccharide units takes place by glycosidic linkage (Figure 1), as takes place in oligosaccharides.

Polysaccharides are also called *glycans*. Their chemical structures may be *linear to highly branched*. The structure may also be *heterogenous, i.e.*, there may be some modification of the monomers (= repeating units) in a polysaccharide unit. Apart from carbohydrate, non-carbohydrate molecules may also remain present in a polysaccharide unit.

The characteristic features of polysacharides are, they are *amorphous* (= not crystalline) and they cannot make *chemical reduction* of copper sulphate, silver nitrate *etc.* Also that, they do not have the property of *osazone* formation. Except cellulose, other polysaccharides are soluble in water and make colloidal solution. They have no sweetness and are almost tasteless. Polysaccharides have great importance in diets as *dietary fibres*, which have been stated under the heading, 2.8 (b).

There are several ways of classifying polysaccharides. When the basic part of a polysacharide polymer is a *pentose sugar* (C_5), it is called a *pentosan* and when the basic part is a *hexose sugar* (C_6), it is called a *hexosan.* Examples of pentosan are araban and xylan *etc.* and those of hexosans are cellulose, glycogen, starch, galactan, mucilage, hemicellulose *etc.*

In accordance with the form of manosaccharide units which constitutes a polysaccharide unit, they are made into two groups. These are, (i) homo-polysaccharides and (ii) hetero-polysaccharides. The difference between these two is, in the former case, only same type of monosaccharide monomers are present, while in the latter case, the monomers are of different types. Important compounds under the homo-polysaccharidegroup are glycogen, starch, inulin, cellulose, pectic substances, chitin *etc.* Under the hetero-polysaccharide group, important compounds are pectido-glycan, proteo-glycan, glyco-protein, muco-protein, glyco-lipid *etc.*

According to *functional importance*, polysaccharides are recognized as *storage-polysaccharides and structural-polysacharides.* The former polysaccharides have the property to store *glucose* for the purpose of using them later (when becomes necessary) and the two common examples are *glycogen* and *starch*, which are actually enriched with glucose molecules. The latter group, *i.e.*, structural polysaccharides give rise to structural construction of cells and body parts of human beings and animals. Some common examples in this group are *arabinoxylans* and *chitin.*

Many chemists classify polysaccharides into *simple form* and *complex form.* Complex polysaccharides are composed of many more monosaccharide units than the simple form. Nutritional value of the complex polysaccharides is known to be higher and dietary fibres come under this group.

2.5.1. Some Important Carbohydrates and their Characteristics

The characteristic features of some of the mono-, di- and poly-saccharides have been stated in the following: Descriptions have however, been made in brief and for detailed knowledge, standard books on organic chemistry and biochemistry should be consulted.

(a) Monosaccharides

About 70 types of monosaccharide compounds are known to exist on Earth. The characteristic features of a few are stated below:

Biose

(i) Glycolic Aldehyde (= Hydroxyacetaldehyde ; Hydroxy ethanal)

This is the *smallest possible* carbohydrate compound known and contains both an aldehyde as well as a hydroxyl group. It may be mentioned that although the compound conforms to the formula of carbohydrate, *i.e.*, $Cn(H_2O)_n$ but the present day chemists usually do not consider it to be a carbohydrate. The biose exists in the biosphere (= life zone, *i.e.*, which covers land and air) as well as in the interstellar spaces (= zone between the stars). It has been identified in the gas and dust particles that exist near the centre of the *milky way.* Taste of the biose is somewhat sweet, shape is crystalline and is readily soluble in water. Its chemical formula is $C_2H_4O_2$, molecular mass is 60.052 g/mol, melting point is 97°C and the boiling point is 131.3°C. The related aldehydes of glycolic aldehyde are, 3-Hydroxybutanal and Lactaldehyde.

Triose

(i) Glyceraldehyde

This is a combination of *glycerol and aldehyde, i.e.*, glycerol with one hydroxymethyl group which is oxidized to an aldehyde. Taste of triose is somewhat sweet and shape is crystalline. It is colourless and is easily soluble in water at the rate of 17 mg in 1 ml. Chemical formula of glyceraldehyde is $C_3H_6O_3$, molecular mass is 90.078 g/mol, melting point is 145°C and boiling point is 150°C at 0.8 mm mercury.

(ii) Dihydroxyacetone

This is formed during conversion of glucose into lactic acid. The triose is known to be the simplest of all hexoses. Dihydroxyacetone is of much physiological importance in its *phosphate ester form, i.e.*, as dihydroxyacetone-phosphate. Taste of the triose is somewhat sweet and cooling and its shape is crystaline. It is hygroscopic with white colour and is soluble in water. It is also soluble in different organic solvents, *e.g.*, ether (= diethyl ether), alcohol (= ethanol) and rectified spirit, acetone and toluene. Chemical formula of the triose is $C_3H_6O_3$. Its melting point is 89°–91°C and boiling point is 217.7°C ± 217.7°C ± 15°C at 760 mm mercury. Dihydroxyacetone has strong reducing property and it is sometimes used to prepare lotion along with erythrulose for colouring of skin. However, use of these chemicals for cosmetic purpose is retricted now.

Tetrose

(i) Erythrose

In having one aldehyde group, the tetrose comes under aldose group. It is colourless, highly soluble in water and its chemical formula is $C_4H_8O_4$, molar mass is 120.104 g/mol and boiling point is 144.07°C. Erythrose was first isolated from rhubarb and was named so for its hue in presence of metallic alkalis. Phosphate ester of the compound, *i.e.*, erythrose – 4 – phosphate formed as an intermediate compound in the pentose-heptose pathway and in the Calvin cycle.

(ii) Threose

For a terminal aldehyde that a ketone group being present in the chain, it is considered a part of aldose monosaccharides. Threose is highly soluble in water. Its molecular mass is 120.104 g/mol and boiling point is 144.07°C.

(iii) Erythrulose

Having one ketone group, it is a ketotetrose as against erythrose and threose which are aldotetroses. It is soluble in water. Chemical formula is $C_4H_8O_4$ and molar mass is 120.104 g/mol. Erythrulose is used in some skin tanning cosmetics, usually in combination with dihydroxyacetone but this is not much approved now.

Pentose

(i) Arabinose

Arabinose has an aldehyde functional group and structurally, it is analogous to D-glyceraldehyde. It has no colour, shape is crystalline and soluble in water (834 g/L at 25°C). Chemical formula of the pentose is $C_5H_{10}O_5$, molar mass is 150.13 g/mol and melting point is 164°C – 165°C. It is used in food. Arabinose exists naturally in sweet basil and many other plants. Gum arabic is very common material in which it mainly exists as L-arabopyranose. Arabinose is sometimes found to be excreted through urine, which is an abnormality of health.

(ii) Xylulose

This pentose sugar of the ketose group is a metabolite of the glucuronic acid. Chemical formula of the pentose is $C_5H_{10}O_5$, molar mass is 150.13 g/mol, melting point is 144°–145°C and the boiling point is 469° ± 45°C at 760 mm mercury.

(iii) Ribose

Ribose sugar is a component of the nucleotide in RNA. The dextro form of this aldopentose is widely observed in nature. It is soluble in water at 100 gram in a litre. Chemical formula of ribose is $C_5H_{10}O_5$, molar mass is 150.13 and melting point is 90°C–95°C. Ribose is sometimes used as a medicine in congestive heart failure and in chronic fatigue syndrome.

(iv) Deoxyribose

This is a constituent of nucleotide in DNA and not in RNA. This is derived from the ribose by the loss of one oxygen atom. In aqueous solution, deoxyribose exists as a mixture of three forms. The pentose is highly soluble in water. Chemical formula of it is $C_5H_{10}O_5$, molecular mass is 134.131 g/mol and melting point is 91°C.

Hexose

(i) Glucose

The term, glucose is derived from the Greek word, which means sweet. Other commercial names of glucose are grape sugar, corn sugar, blood sugar *etc.* Among all carbohydrates, *most abundant is glucose.* It is actually synthesized in the plant cells following the reaction in plants, which is known as *photosynthesis,* in which the green pigment, known as *chlorophyll* act as energy transformer and enable the cells to use *light as the source of energy.* The overall reaction of the process is as follows:

$$6CO_2 + 6H_2O \xrightarrow[\text{673 kcl}]{\text{Light}} C_6H_{12}O_6 + 6O_2$$

The human beings and animals get the glucose when they take food obtained from plants or other animals who have obtained glucose by the intake of plant-source food. Glucose is also obtained from non-carbohydrate source, such as lactic acid, glycerol, propionic acid *etc.*, which is known as *gluconeogenesis.* This takes place in the liver and also kidney and the process may be active when intake of sugar from food is very low, particularly in the central nervous system, red blood cells *etc.*, in which organs requirement of glucose is higher.

Glucose is colourless, crystalline, considerably sweet in taste and is highly soluble in water. Chemical formula of this hexose sugar is $C_6H_{12}O_6$, molecular mass is 180.156 g/mol, melting point is 146°C (alpha – D – gluose) and 150°C (beta – D – glucose) and boiling point is 150° ± 2°C. It remains as a straight-chain compound or in various ring forms. The isomeric forms of the hexose are alpha and beta glucose.

The optical rotation of glucose is + 112° just after solution of it is made in water. However, when the solution is allowed to stand for sometime, the optical rotation comes down to + 52.5°. This is known as *mutarotation.* But when the solution is made after recrystallization from boiling pyridine, optical rotation comes to +19°, instead of 112°. After standing for sometime, optical rotation of the solution is however, increased, *i.e.*, mutarotated and comes to + 52.5°. This mutarotation phenomenon indicates that in solution form of glucose, one form changes to other form and this continues until an *equilibrium* is reached.

Glucose has the unique property in that, among all sugars, it is the *most accesible* form *of energy in the body.* In cellular respiration, it is oxidized and liberates energy in the form of ATP. It has a great importance in carbohydrate metabolism which takes place on esterification with phosphoric acid. Naturally occurring glucose is in *dextro*-form and the *laevo*-form is produced synthetically.

Glucose occurs in nature as a disaccharide, *e.g.*, in sucrose, lactose, maltose *etc.*, as a polysaccharide in the form of starch and glycogen and also exists in combination, *i.e.*, conjugation with protein as in glycoprotein. From these compounds, it is released on hydrolysis when there is requirement.

Grape berries and many fruits like litchi, sapota, date, jackfruit, oranges *etc.* are rich or fairly rich sources of glucose. Juices tapped from wild date-palm, palmyra-palm *etc.* contain very high amount of glucose. In sugarcane stalks, the top portions are rich in glucose while the lower parts contain high amount of sucrose.

Glucose is the *highly potent sugar in providing energy* in the body and hence, is nicknamed as energizer in games and sports. In the blood plasma of human beings, it remains in approximately 0.1 per cent and the *insulin hormone* released by the pancreas tends to maintain the concentration in the plasma. Glucose is used as the first medicine when instant energy in the body is required. However, more than optimum requirement in the blood plasma (= hyperglycaemia), known as diabetes (type – 2) is a physiological abnormality. Similarly, low glucose in the plasma (= hypoglycaemia) stands in the way to carry on physical activity and brings about serious weakness if it is too low.

(ii) Galactose (= Milk sugar; Gal)

This hexose sugar is a component of lactose which is a disaccharide of galactose and glucose. Galactose *is present in milk* and does not occur in free form. When lactose molecule is hydrolyzed, galactose and glucose molecules are liberated, *i.e.*, obtained. Polymer form of galactose is galactan and when this polymer is hydrolyzed, that is, enzymatically split, galactose monomers are liberated. Galactose exists in open-chain and in cyclic forms. It tastes sweet and is soluble in water at the rate of 650 gram in a litre at a temperature of 20°C. Chemical formula of it is $C_6H_{12}O_6$ and its molar mass is 180.16 g/mol. Melting point of galactose is 167°C and boiling point is 527.1° ± 50 °C at 760 mm mercury. It can be derived from sugarbeet. In the human body, it is *present in much amount in brain*. Galactose is a constituent of *cerebroside*, which is a fatty substance in the nervous tissue. The monosaccharide is readily absorbed in the gut. In the liver, it is *converted to glucose* and is stored as glycogen.

(iii) Fructose

The term, fructose is derived from the Latin word, *fructus*, which means fruit and this points out the fact that *fruits are rich sources* of this simple ketonic monosaccharide. Glucose, fructose and galactose are in fact, the three dietary sugars which are absorbed from the blood directly in course of digestion. Like glucose, fructose has a great importance in carbohydrate metabolism which takes place on *esterification* with *phosphoric acid*. Fructose is *sweetest of all carbohydrates*. Pure and dry fructose is white in colour, has no odour and is crystalline. It is the *most water-soluble* of all sugars and unlike other sugars, it is soluble in hot alcohol (= ethanol). Its chemical formula is $C_6H_{12}O_6$, molar mass is 180.156 g/mol, melting point is 103°C, boiling point is 440°C and optical rotation is – 92°C.

Fructose is present in many fruits, palm juices, root tubers of dahlia plant, sugarcane juice, sugarbeet, maize *etc.* and in these materials, it remains bonded to the isomer, glucose so as to form the sucrose which is disaccharide of glucose and fructose. However, *honey is the richest source* of this ketose. Honey contains fructose, glucose and sucrose but fructose is in much greater proportion and hence, it is sweet. (Honey also contains pollen grains of flowers and insect excreta carried by the bees. In containing pollen grains, it may produce allergy in some persons). Fructose is a constituent of the polysaccharide, *inulin* and it can be obtained from this by hydrolysis. In the human body, it does not normally occur in free state, because it is readily *converted into glucose in the liver and intestine.* Increase of blood is however, low as compared to sucrose and glucose. Nevertheless, too much intake of this ketose sugar is not desirable as it leads to insulin resistance, *i.e.*, type – 2 diabetes, obesity, increase of cholesterol and triglyceride (fat) in the blood plasma.

(iv) Mannose

Mannose, a C – 2 epimer of glucose has its importance as a constituent of some animal proteins. It has significant role in metabolism in regard to glycosylation of some proteins. Its glucoside form (= mannoside) has also role in nervous system. Mannose is white in colour, crystalline and soluble in water at 50 mg/ml. Its chemical formula is $C_6H_{12}O_6$, molecular mass is 180.156 g/mol, melting point is 133°–140°C and boiling point is 293.96°C approximately. Mannose does not occur free in nature but occurs abundantly in its condensation product, mannan from which it is obtained by boiling with acid. It is also found in some nuts (ivory nut) and softwood tissues in plants. (Combination of galactose and mannose is termed as galacto-mannan).

Heptose

(i) Sedoheptulose

This 7 carbon-atom ketoheptose monosaccharide is among few heptoses found in nature. In photosynthetic and respiratory pathways, *this heptose monosaccharide has a significant role* and is formed as an ester of phosphoric acid in the metabolic pathway of pentose phosphate by hexose phosphate shunt as an intermediate compound. Heptose is soluble in water. Its chemical formula is $C_7H_{14}O_7$, molecular mass is 210.182 g/mol and boiling point is 628° ± 55°C at 760 mm mercury. Sedoheptulose occurs in apple, apricot and some other fruits. The compound reduces pro-inflammatory markers and C–reactive protein and thereby reduces some inflammations in the body.

(b) Oligosaccharides

Disaccharide

(i) Sucrose

The word, sucrose is derived from the French word, *sucre*, which means sugar. We are familiar with common sugar and sucrose is the purified form of the common sugar. It is the most abundant of all disaccharides. It has been stated under the heading, 2.5.1 that a molecule of the disaccharide sucrose is formed by the condensation of one molecule, each of D–glucose and D–fructose by *glycosidic bond* (Heading 2.5.1), in which one molecule of water is eliminated. Sucrose is neither an aldehyde nor a ketone and hence, *cannot make chemical reduction* of copper sulphate, silver nitrate *etc.* It does not form osazone also.

Synthesis of sucrose takes place by several ways as follows:

(a) Glucose–1–phosphate + fructose = Sucrose.

(b) Uridine dilophosphate glucose (UDPG) + fructose = Sucrose.

(c) UDPG + 6–phosphate = Sucrose. Normally however, synthesis takes place by the following reaction. Fructose – furanose 6–phosphate + glucose – pyranose 6–phosphate. This reaction takes place in presence of condensing enzyme and the uridine triphosphate.

As regard optical density, sucrose solution is strongly *dextrorotatory.* But when it is hydrolyzed, the resulting mixture of glucose and fructose is found to be laevorotatory. This is because, laevorotation of fructose is greater than dextrorotation of glucose. Hence, the mixture is called *invert sugar*. Accordingly, the enzyme which hydrolyzes the sucrose molecule is called invertase. (However, the enzyme is more commonly termed as *sucrase* than invertase). Sucrose gives high energy in the body. When it is taken by mouth, it is hydrolyzed in the intestine and gives rise to *glucose* and *fructose*. Glucose molecules are oxidized in the cells and give rise to energy in the form of ATP. Fructose molecules are however, converted to glucose first and then glucose molecules are oxidized in the cells to liberate energy. Thus, a sucrose molecule ultimately gives rise to two molecules of glucose and becomes rich source of energy. As regard release of energy, sucrose may therefore, seem to be more effective than glucose. But when sugar is taken in mouth, it takes some time to reach intestine and also takes some time there to get hydrolyzed. Therefore, if a person shows symptoms of sudden fall of blood sugar (hypoglycaemia), D–glucose is a better option to give rather than sucrose to get energy at a quicker time.

Pure sucrose is solid, white in colour, crystalline and is highly soluble in water (~2000g/L at 25°C). Its taste in very sweet but less sweet than fructose. Its chemical formula is $C_{12}H_{22}O_{11}$, molar mass is 342.30 g/mol, melting point is 186°C and density is 1.587 g/cm^3. In nature, high amount of sugar is present in sugarcane juice which is 10 - 13 per cent and in sugarbeet root, in which it is 12 - 18 per cent. High amount of sugar is present in palm juices and many fruits, *e.g.*, date, sapota, grape berries, cranberries, pineapple, mango *etc*. However, around the world, common sugar of commerce is recovered to about 56 and 44 per cent respectively from sugarcane juice and sugarbeet root extract. (In India, sugarcane is most commonly used).

(ii) Lactose

The word, lactose is derived from the Latin word, *lac* (= lactis) which means milk. One molecule of this disaccharide is formed by condensation of one molecule, each of galactose and glucose. The galactose and glucose molecules are condensed together by glycosidic linkage (Figure 1) as in case of sucrose. Galactose has only the beta-pyranose form but glucose may be either alpha- or beta-pyranose.

Lactose is present *only in milk and hence, it is called milk sugar*. It is present in milk of all mammals. In milk, it is present from 4 to 7 per cent. Lactose can make chemical reduction of compounds like copper sulphate, silver nitrate *etc*. It also forms osazone. These indicate that lactose is an *aldose*. Chemically, lactose is grouped under alpha–, beta–, gamma– and delta– isomers, in accordacne with position of carboxy atoms.

Pure form of lactose is white in colour, solid and crystalline. It is mildly sweet in taste, soluble in water at 195 g/litre. Chemical formula of lactose is $C_{12}H_{22}O_{11}$, molar mass is 342.3 g/mol, melting point is 202.8°C and the density is 1.525 g/cm^3.

The curd, a milk product made in India is prepared by souring milk in cold condition. *Plentiful bacteria infect and grow vigorously in milk when it is soured* and these bacteria produce lactic acid, butyric acid and other organic acids in the milk. Due to that, curd is set from the milk. Addition (= inoculation) of desirable quality of bacterial culture, known as *starter* produces curd of high quality and medicinal value. If the curd becomes necessary to be sweetened (= dulcified), it is advisable that the sugar should be mixed after the curd is formed and should not be mixed with milk before souring as the curd of the latter type is of less nutritive and medicinal value.

A characteristic feature of lactose is, when it goes to the body of some people (but not everyone), some undesirable symptoms develop in them. These symptoms (better to say, syndromes) are gastro-intestinal disturbances, bloating, diarrhoea, cramps *etc.* This is medically referred to as *lactose intolerance.* The reason for this abnormality is hereditary, *i.e.*, genetic. This is commonly observed in the Asian people and also in the prematurely born infants. In India, about 20 per cent of people have this abnormality. Those people who show lactose intolerance syndrome cannot digest high amount of milk but which is easily digested by the European people without any adverse symptom.

Physiologically, the reason for lactose intolerance is an enzyme. This enzyme is called *lactase* (= beta–D–galactosidase) which is essential for absorption of lactose from the intestine. Those people who have lactose intolerance abnormality are in fact, deficient of the lactase enzyme. This enzyme breaks down the lactose in the small intestine and hence, only the lactose-free milk goes from the small to the large intestine and thus, it is easily absorbed. Nowadays, lactose-free milk is available in the market. Therefore, those people who have the genetic disorder of lactose intolerance should consume lactose-free milk. It may be pointed out in this connection that almost in every person, production of the enzyme, lactose falls with age and if necessary, the elderly persons should also consume lactose-free milk, if there is necessity.

It is brought to notice in this connection that sometimes lactose intolerance is *confused with milk allergy*. Lactose intolerance is genetic (= hereditary) phenomenon when lactose sugar in the milk is not digested (= broken down) due to lack or insufficient production of the lactase enzyme. Milk allergy is also a genetic phenomenon in which some proteins present in the milk (act as allergens) and causes allergy. Symptoms of milk allergy are skin rashes, oedema, severe itching, sneezing *etc.* Milk allergy begins mostly in childhood.

(iii) Maltose

The word, *malt* is derived from the word, *mealt.* By malt is meant the germinated grains of cereals and particularly, *barley*. In the distillery industries, it is much used and it is prepared by germinating the grains of barley under controlled environment after which, the germinated barley grains are heated and dried.

Malt is rich in proteins and many enzymes and among enzymes, *diastase* is of great importance. Diastase hydrolyzes starch and converts maltose to glucose. The extract which is made from malt is called *wort* and from this, maltose is made.

Maltose does not occur is free form. Maltose molecule is formed by condensation of *glucopyranose* by glycosidic bond. As maltose has aldehyde group, it can chemically reduce coper sulphate, silver nitrate *etc.* and forms *osazone.* In having glucose, the disaccharide, lactose and maltose provide energy in the body on hydrolysis. Isomaltose is an isomer of maltose but instead of bond in the 1, 4 – position, it is in the 1, 6 – position. In pure form, maltose is white, solid, crystalline and shows mutarotation in aqueous solution. Its chemical formula is $C_{12}H_{22}O_{11}$, molar mass is 342.30 g/mol, solubility in water is 1.08 g/ml at 20°C, melting point is 160°–165°C and the density is 1.54 g/cm^3.

Trisaccharides

(i) Maltotriose

This trisaccharide molecule is formed by condensation of three molecules of glucose which are linked by 1, 4 – glycosidic linkage. This is produced by the digestive enzyme, alpha-amylase, which is a common enzyme in saliva and hydrolyzes starch. It is soluble in water. Chemical formula of maltotriose is $C_{18}H_{32}O_{16}$ and molar mass is 504.438 g/mol. Maltotriose is the shortest chain of all oligosaccharides.

(c) Polysaccharides

(i) Araban

The pentosan, araban is a polysaccharide of the pentose sugar, arabinose. This polymer is relatively short-chained and is composed of L–arabinose monomer units and these are joined (= linked, bonded) together by alpha–1, 5 linkage. The polysaccharide occurs in the cells of plants and most common is *mucilage*. Araban is present in some fruits also, *e.g.*, plum, peach, cherry *etc.*

(ii) Xylan

Apart from araban, the other pentosan polysaccharides of importance is xylan which is also a short chain polymer, made up of D-xylose units as the monomers. The monomer units are condensed together by beta – 1, 4 linkage. It occurs in many aquatic (water) plants and as mucilage in the cacti. Xylan also occurs in straw, bran and the shells of apricot seeds. Araban and xylan are most common pentosan polysaccharides in plants and these apart, fucosan may also be named which mostly occurs in the cells of many aquatic plants and algae.

(iii) Glycogen

Glycogen, the unique storage material of glucose molecules in the body is derived from the Greek word, *glyco,* which means sweet or materials that are related to sweetness. This polysaccharide is present only in the body of human beings and animals and not in any plant. This glycan, *i.e.*, polysacccharide was first disovered (1857) in liver by claude Bernard. Shortly after that, A. Sanson observed its existence in the *muscle* tissue also.

Function of glycogen is to *store glucose molecules in the body as reserve materials,* such that the stored glucose molecules could be made use of when there would be necessity. In this sense, glycogen is *analogous to the hexosan polysaccharide, i.e., starch* which is present in the plants. Starch has the same function, *i.e.*, it stores plenty of glucose molecules for the purpose of using them in case of necessity. Hence, glycogen, the storehouse of glucose in the human beings and animals is sometimes nick-named as "animal starch" and likewise, starch is nick-named as "plant glycogen."

In the body, glycogen is abundantly present in the *liver* and *muscles* and the *hormones,* insulin and cortisol *facilitate in the process of storage.* In the liver, glycogen is present in 3 – 7 per cent and in the muscles, 0.5 – 1 per cent. In a healthy and well-nourished adult, the total amount of glycogen that may be stored in the liver may be about 200 gram and in the muscles, the total storage amount may be about 300 gram. Higher amount in the muscles is due to the fact that the total mass of muscles in the body when considered together would evidently be much higher in the muscles than in the liver. Glycogen is stored in the liver and muscles in *moist form.* That is, for one part of glycogen, 3 – 4 parts of water remains mixed to give it a moist appearance. The glycogen present in the liver is easily converted by enzymes to glucose molecules and the glucose molecules are also easily converted to glycogen by enzymatic action.

It may be brought to notice in this connection that although glycogen is in abundance in the liver and muscles, it is present in other parts also, *e.g.*, in brain, stomach, kidney, white blood corpuscles and also in the uterus during pregnancy. However, as compared to liver and muscles, amount that is present in other organs is significantly lower. Another interesting feature is, glycogen has its existence in some species of fungi and bacteria. As a rule, glycogen is not present in plants but interestingly, a compound similar to this compound has been reported to exist in a plant, which is the *Golden Bantam* bantam variety of sweet corn.

Glycogen is formed by condensation of very large number of glucose molecules which are the monomers, *i.e.*, the individual units. Its chemical structure is formed of *branched chains.* The length of chain varies from 8 to 12 glucose units. The glucose units (= molecules) are linked together to the chain by alpha-1– 4 – glycosidic bonds from one molecule of glucose to the other. The basic chains which are linear are branched off by alpha- 1–6–glycosidic bonds. This takes place between the first molecule of glucose of new branch and a glucose molecule on the stem chain. Thus, the chemical structure of a glycogen molecule is similar to that of *amylopectin* which is one of the two compounds that combine together to form a starch molecule and this has been elaborated under the next heading on starch. However, unlike amylopecticn, side-chains of glycogen are more extensively branched.

Glycogen exists in the form of *granules* (= also known as grains) and every glycogen granule has in its core, a *glycogenin protein.* Colour of glycogen is white and its molecular mass ranges from 105 to 108. It is soluble in water and gives an opalescent solution. When reacted with iodine, it gives reddish colour.

Glycogen is most commonly formed from glucose molecules as the source material and the process is known as *glycogenesis*. In the synthesis, uridine triphosphate (UTP) which provides energy reacts with glucose-1-phosphate, forming UDL-glucose, which is catalyzed by the enzyme, UTP-glucose-1- phosphate uridylyltransferase. It is synthesized from the monomers of UDP - glucose by the glycogenin protein which is a homodimer. This protein has two amino acid anchors of tyrosine for the reducing end of glycogen. When 7 - 8 glucose units are attached to a tyrosine residue, the glycogen synthase enzyme elongates the glycogen chain length using UDP- glucose and attaches the alpha- 1-4-linked glucose molecule to the reducing end of the chain of glycogen. The enzyme responsible for branching glycogen catalyzes the transfer of a terminal part of several glucose units from a non-reducing end to the C-6 hydroxyl group of a glucose molecule deeper into the glycogen molecule. The enzyme for branching acts on a branch which has 11 or more units the enzyme may transfer to the same or other chains of glucose molecules. Glycogen may be synthesized from *non-carbohydrate source* also such as fat, amino acids *etc.*, and the process is known as *glyconeogenesis.*

Hydrolysis, *i.e.*, catabolism of glycogen takes place by the enzyme, glycogenase in the liver and this is known as *glycogeneolysis*. By this breakdown of glycogen molecules, glucose molecules are released and this takes place when the body needs glucose for energy. The glucose molecules released by hydrolysis of the glycogen in the liver go to the blood stream directly. A hormone known as *glucagon* which is produced by the pancreas takes part in the releasing glucose molecules from the glycogen. Glycogen which is stored in the muscles is however, broken down at first into glucose-6-phosphate and then enters glycolytic pathway to liberate glucose molecules. Synthesis and breakdown (hydrolyis) go on simultaneaously.

This glycan has a great significance in the body as it stores high amont of glucose moleules for use in necessity and thereby, excess amont of glucose, if any is saved. However, excess storage of glycogen may become a problem. This may happen if a person consumes too much sugar or starch. Accumulation of very high glycogen may also be a heritable character. Such abnormal high accumulation of glycogen particularly in the liver is a health hazard and recognized as *glycogen storage disease.*

(iv) Starch

Starch, other name of which is *amylum* is derived from the German word, *starke* and the Greek word, *amylon*. It is a polysaccharide which is formed by the condensation of very large number of glucose molecules. This glycan is synthesized only *in plants, not in any animal but is analogous to glycogen which is present only in the body of human beings and animals* and not in any plant. It is largely present in some species of plants, notably cereals and millets (= minor cereals), *e.g.*, rice, wheat, maize, barley, sorghum, pearl millet *etc.*, in which it is 65-85 per cent and is also abundant in root and stem tubers and seeds where it is 19-35 per cent. Similar to glycogen in animals, significance of starch produtcion in plants lies in *storage of glucose*. That is, extra glucose molecules which are not utilized in a plant body are *stored in it in the form of starch*. This is done with the idea that the glucose molecules could be utilized by them when there will be necessity.

Synthesis of starch is done in the leaves of the plants during day time. It is stored as *granules, i.e., grains*. The granules are translocated to the storage organs. In synthesis, glucose–1–phosphate is converted to ADP–glucose by the enzyme, glucose–1–phosphate adenyltransferase as the first step. (In this connection, it may however, be pointed out that all plant species do not store excess glucose molecules in the form of starch. Some species of plants, *e.g.*, those of the botanical family, compositae, such as dahlia, Jerusalem artichoke *etc.* could be mentioned which store excess glucose in their tuberous roots in the form of the polysaccharide, *inulin* and not as starch). Starch has a great significance to the human beings as it is the *principal form of carbohydrate in their food and is present in very high amount in the staple food.* It has no taste or odour and is white in colour.

When seen through microscope, all starch granules do not look to be same, *i.e., their shape differs in accordance with plant species from which it has been obtained.* During day time or under illuminated condition, the granules become amorphous and contain different amount of moisture than when they are formed at night or in darkness.

An important feature should be kept in mind in respect of starch. That is, what we call starch is chemically not a one substance. As a matter of fact, *starch is a mixture ot two carbohydrates*. These two carbohydrates are called *amylose* and *amylopectin*. When the amylose and amylopectin molecules are combined (= condensed) together by enzymes, the starch molecule is formed. Thus, starch molecule is actually a *combined product of amylose and amylopectin* molecules.

The carbohydrate compounds, amylose and amylopectin are not same. The basic difference between these two types of carbohydrates may be stated as follows:

Amylose is a long, *unbranched chain* of glucose molecules which are linked together by alpha-1–4–glycoside and in this process, one molecule of water is eliminated. The linkage in amylose molecules occurs as in maltose and accordingly, when it is split (= hydrolyzed) by the enzyme, amylase, it *liberates maltose* molecules. However, it may be brought to notice in this connection that although amylose was formerly considered as a chain that is completely unbranched, it is not so. In fact, some of the molecules, although few that are contained in it have some branch points. One molecule of amylose consists of 500–5000 molecules of glucose. Molecular mass of amylose ranges from 10^5 to 10^6. From X-ray diffraction, it appears to be crystalline. As regard solubility, amylose is soluble in water but the solution on standing becomes turbid due to precipitation of amylose and this happens due to the process known as *retrogradation*. Amylose forms complexes with fatty acids and alcohols of low molecular weight. When reacted with iodine, it gives blue colouration.

On the other hand, the carbohydrate, amylopectin is a *very large and highly branched* carbohydrate molecule. It has similar chain of alpha-1–4 glycosidic linkage like amylose but apart from this, *there are many side-chains* which are attached to the basic chain by alpha-1–6–glycosidic linkage. A molecule of amylopectin may be composed of 0.5 to 5 lakh glucose molecules. Amylopectin is amorphous, *i.e.*, not crystalline and the aqueous solution of it is stable, in other words, does not retrograde to become turbid when allowed to stand. When reacted with iodine, it gives purple colouration.

It needs mentioning that the amylose and the amylopectin content of starch belonging to all species of plants are not same. For example, in the starch which is obtained from rice, wheat, maize, sorghum, potato and banana, amylose is present to the extent of 25, 24, 28, 22, 22 and 21 per cent and amylopectin to about 75, 76, 72, 78 and 79 per cent respectively. In general, however, in the starch derived from most plant species, amylose content ranges from 15 to 30 per cent while amylopectin content is 70 to 80 per cent.

Hydrolysis of starch is done by the enzyme, *amylase* (or diastase). This is a complex enzyme of three or more enzymes, which are alpha-amylase, beta-amylase and *R*–enzyme. The alpha-amylase (Enzyme Commission No. 3.2.1.1.) splits 1–4 glycosidic linkages in all places of amylose and amylopectin and the beta-amylase (EC No. 3.2.1.2) also splits 1– 4 glycosidic linkage but splits maltose molecules from terminal parts of amylose or amylopectin chains. The *R* –enzyme (EC No. 3.2.1.41) hydrolyzes 1–6 linkage of the amylopectin molecules. Starch is also hydrolyzed by hot, dilute mineral acids to liberate glucose molecules. When cooked form of starch is taken as food, it is completely digested (= hydrolyzed) in the gastro-intestinal tract, partly by the amylase of the saliva and mostly by the amylase of the pancreas. On digestion, glucose molecules are liberated and these are used as source of energy in the body.

(v) Dextrin

These hexosan polysaccharides do not occur naturally and are formed as intermediate products during hydrolysis of starch or glycogen. The intermediate products are named according to their *colour formation* when reacted with iodine. These are amylodextrins, erythrodextrins and achrodextrins which respectively give blue, red and no colour on reaction with iodine. Dextrins are produced on heating starch also under dry and acidic condition. They belong to low molecular mass carbohydrates and are mildly sweet in taste.

(vi) Inulin

This glycan is formed by condensation of a large number of D–fructose molecules. The fructose units are joined in straight chain by beta-2–1 glycosidic linkage. Inulin is crystalline and is soluble in hot water. It is white in colour and does not give any colour when reacted with iodine. In the root tubers of dahlia and some other plants, fructose molecules are stored in the form of inulin.

(vii) Galactan

This hexosan polysaccharide is formed by condensation of galactose molecules. The galactose units are joined by alpha-1–3 or alpha-1–6 glycosidic linkage. It is present in soybean seeds and in wood of some species of trees.

(viii) Mannan

This hexosan (polysaccharide) is formed by condensation of mannose units, joined by alpha-1–4 glycosidic linkage. The principal types are glucomannans and galactomannans. Mannan is present in coffee, custard apple and other plants. It is also present in yeast and bacteria. Mannan is of great use in food industries particularly as a thickening agent.

(ix) Cellulose

Cellulose is a polysaccharide (= glycan) which is known to be *most abundantly* present on Earth among all types of carbohydrates. It is estimated that about 1015 kg of cellulose is regularly synthesized and destroyed in the Earth. The polysaccharide was first discovered by the French chemist, Payen in 1838. Cellulose is *present only in plants and never in any animal.* In the plant, it is present in the *cell-wall* along with other polysaccharides, *e.g.*, pectic substances, lignin, hemicellulose *etc.* and these combinations provide *structural support* to the protoplasm. High cellulose is found in cotton fibre where it is about 90 per cent. It is also high in the wood of trees where it is 40–50 per cent. Some food-stuffs contain appreciable amount of cellulose, *e.g.*, apple, apricot, asparagus, beans, oat, mushroom, orange, onion, wheat husk, prune fruits *etc.*

Chemically, cellulose is formed by condensation of 1.5–5 lakh glucose residues (= units, molecules) and this remains as an *unbranched chain* of glucose molecules. *It differs from starch in the mode of linkage between glucose molecules.* That is, in the starch molecule, the alpha-glucose units are joined by 1–4–glycosidic linkage but in cellulose, instead of alpha units, beta-glucose units are only linked. Due to such chemical structure, cellulose is not acted upon by amylase enzyme in the gastro-intestinal tract. Hence, it is *not digested* and pass out through faeces as such without undergoing any change, *i.e.*, as undigested material. Digestion of cellulose is observed in carnivorous animals, termites *etc.* but in fact, the bacteria (and may be fungi) which are present in their gastro-intestinal tract digest the cellulose and thus, they have some *symbiotic relationship* with these animals. The cellulose is although a linear polymer of glucose molecules, under electron microscrope, the glucose molecules are found to be clumped together into long threads, which are called *microfibrils*. Cellulose is fibrous, tough, has no taste or odour, white in colour and is hydrophilic. It possesses both crystalline and amorphous character and is much crystalline than starch. It is *not soluble in water or in any organic solvent* but may come into solution in ammoniacal copper hydroxide which is known as Schweitzer's reagent. It is slowly hydrolyzed by concentrated sulphuric acid but dilute sulphuric acid or dilute sodium hydroxide causes it to swell.

Although cellulose has a protective role to save the protoplasm of the cell against possible damage, it is of significance to the human beings also. It is not digested in the body and provides no nutrition but *stimulates peristalsis* and is of great help in evacuation, preventing many diseases and countering obesity. Cellulose is however, of great industrial value. In the making of paper, cellophane, textiles, some types of glass fibres, it is very much used. Cellulose has a great role in wood industry as well.

(d) Complex Polysaccharides

These compounds are *present only in plants and not in any animal.* Some polysaccharides belonging to this group are discussed below: Their formulae are much more complex.

(i) Gum

Gum is produced by some plants as a by-product, either naturally or under stress. It is exudated slowly from leaves, roots but most commonly from bark. Some of the common gums are gum arabic, cherry gum, damson gum, mesquite gum *etc.* Exudation of gum from the surface of bark is a common symptom in citruses and some other trees when they are infected by fungi and notably, when infected by the species belonging to the genus, *Phytophthora* of the Phycomycetes class. Most common symptoms by *Phytophtora* infection are exudation (= discharge) of gum-like semi-transparent substance (= gummy exudation) from the bark. Later, the oozed out gum becomes somewhat hardened. The condition, *i.e.*, the symptom is called *gummosis*. Gummosis not only occurs due to pathogenic (= fungal) infection but also due to stress of other kinds. For example, it is found that trees of wild cherry exudate gum in stress due to low temperature, *i.e.*, when temperature goes very low in winter season. Gummy exudation is found in many aquatic plants also on the outer walls of them.

Gum is a sticky, amorphous (= not crystalline), transluscent (= not fully transparent or opaque) material and becomes softened on taking up water, *i.e.*, when dipped in water.

Chemically, gum is a *branched chain* in structure, which is formed by condensation of various pentoses, hexoses and acids which are derived from sugars. When hydrolyzed, gum gives rise to sugars, *viz*, arabinose, rhamnose, futose (= methyl pentose), besides, mannose, galactose, galacturonic acid, glucoronic acid *etc.* Physiological role of gum in plants is not clearly known. Gum is however, used in some industries, *e.g.*, as a thickening agent in food industries, glue, adhesive, textile and in making emulsifiers.

(ii) Mucilage

Mucilage is a gluey substance produced by a large number of plants to a greater or a lesser extent. It is similar to gum in many properties, such as both are insoluble in alcohol but liquefy and swell in water, transluscent and amorphous.

Main difference is, mucilage does not dissolve in water like gum. This glycan (= polysaccharide) has some significance in plants in the storage of water and perhaps for this reason, it is abundantly present in the cacti which are largely grown in the deserts. High mucilage is also found in the bark of elm, seed-coats of flax, mustard, isapgol (*plantago*), many legumes *etc.* Agar (*agar-agar*) is a mucilage present in the sea weed, *Rhodophyta* species. Some micro-organisms also store mucilage in their body. Interestingly, they use it for their movement, *i.e.*, they go on moving opposite to the secretion of mucilage. Mucilage has many medicinal properties. It acts as a protective film in the lumen and relieves irritation of the mucous membrane and is also helpful in constipation. In many food industries also, mucilage is used.

(iii) Hemicellulose

This polysaccharide is also known as *polyose* and this comprises L – arabinose, D–galactose, D–glucose, D–mannose and D–xylose which are joined together in different combinations by glycosidic linkages. Hemicelluloses are not termed as *alkali-soluble polysaccharides* and these are in close association with cellulose. They are one of the important components of cell-wall in plants and are composed of glycosidic chains having molecules of both uronic acid and pentose.

Hemicelluloses differ from the cellulose in the chain length. While they consist of shorter chains of 500–3000 glucose molecules, in cellulose, the chains are longer and consist of 7000–15000 glucose molecules. They differ from pectic substances in becoming soluble in alkali and not in water. Large amount of hemicelluloses is present in the wood of trees. They are also abundant in the endosperm of many seeds. In some cereals, hemicelluloses accumulate when there is stress of water.

(iv) Pectic Substances

These are calcium or magnesium esters of galacturonic acid and are present in the middle lamellae and also in the primary cell-walls of plants. Four types of pectic compounds exist and these are, *protopectin, pectinic acid, pectin and pectic acid.* Characteristic features of these compounds are as follows:

1. *Protopectin* : These are not soluble in water and are recognized as parent pectic compounds. On restricted hydrolysis, protopectins liberate pectin and pectinic acid molucules.
2. *Pectinic acid* : These are colloidal form of polygalacturonic esters which contain a negligible proportion of methyl esters.
3. *Pectin* : These are water-soluble pectinic acids of varying methyl esters and degree of methylation. That is, several of the carboxyl groups of the galacturonic acids in these compounds remain esterified with methyl groups.
4. *Pectic acid* : These are long-chain polymers of galactose molecules with several molecules of galacturonic acid and arabinose.

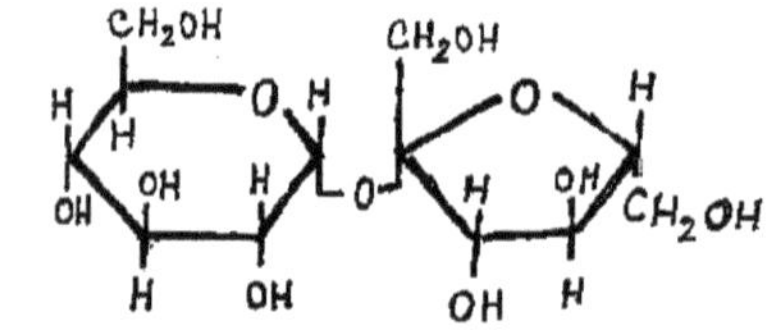

Glucopyranose Fructopyranose

Sucrose

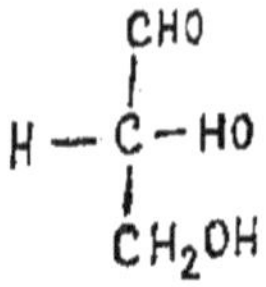

D-Glyceraldehyde

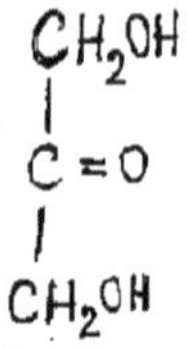

Dihydroxyacetone

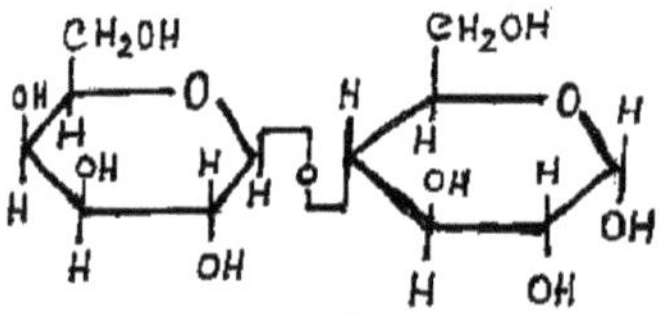

Galactose Glucose

Lactose

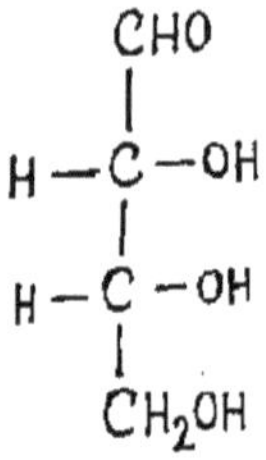

D-Erythrose

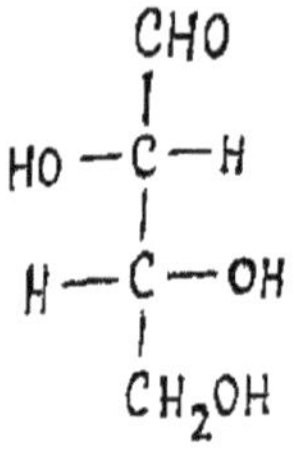

D-Threose

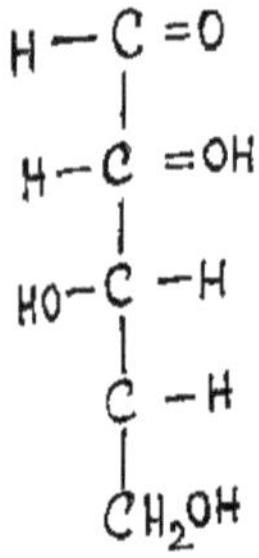

L-Arabinose

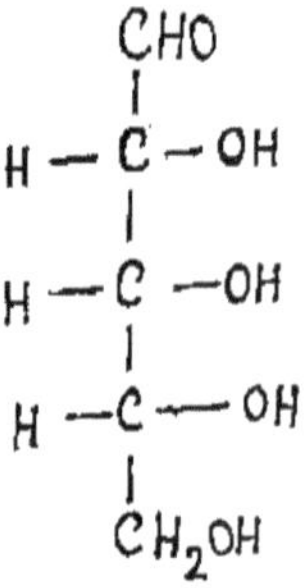

D-Ribose

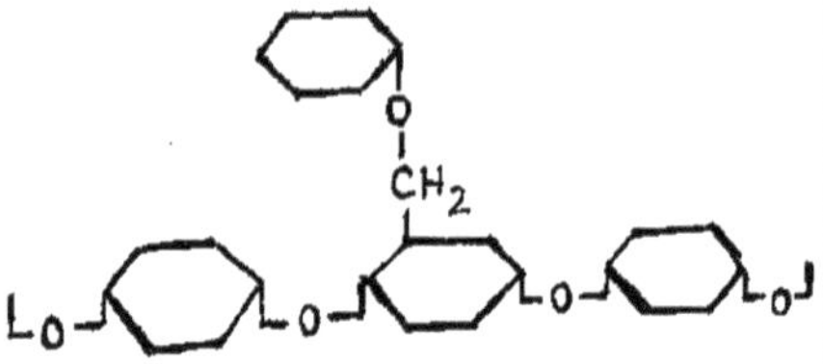

Glycogen

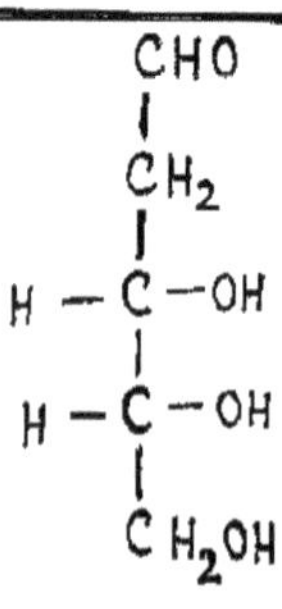

D-Deoxyribose

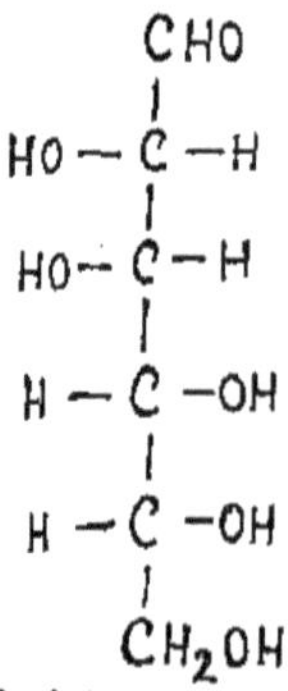

D-Mannose

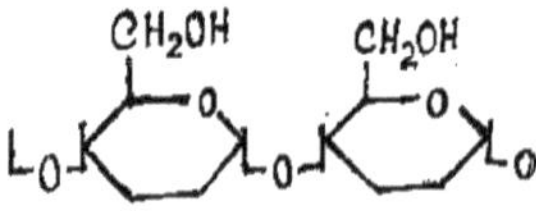

Amylose

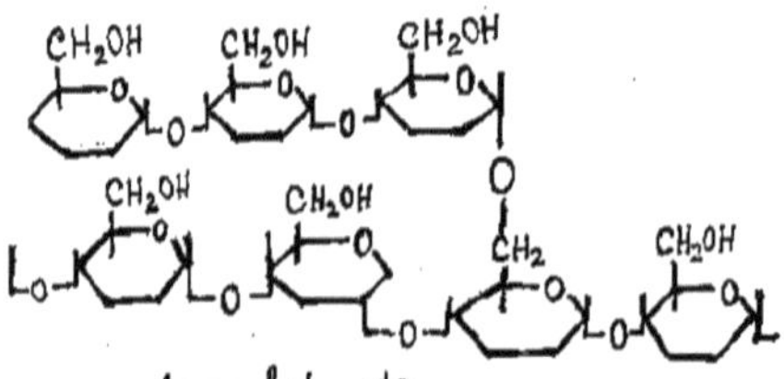

Amylopectin

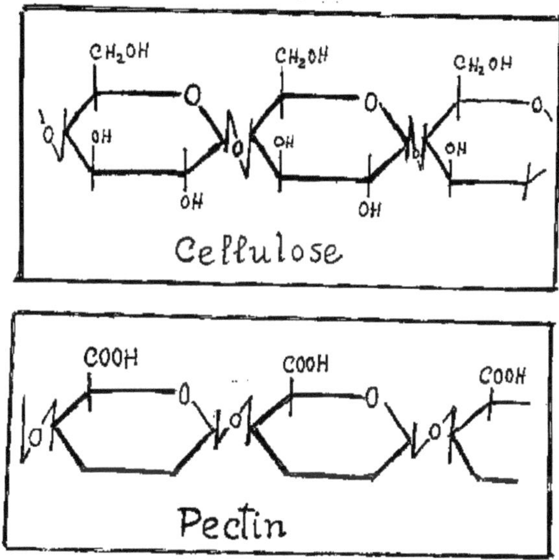

Figure 2. Structures of some Monosaccharides, Oligosaccharides and Polysaccharides.

By the action of pectic enzymes, the parent compound, protopectin undergoes changes to pectinic acid and subsequently, the pectinic acid enzymatically changes to pectin and ultimately, pectic acid. By this enzymatic conversion, *chain length of the pectic compounds shortens, their degree of methylation decreases* and *rate of solubility in water increases.*

In fruits, it is found that they are hard in texture when they are young and immature. At this stage, protopectin remains abundant in them. As the solubility of these compounds is very low, they hold the adjoining cells firmly together and fruits are very hard in texture. With progress of growth of the fruits, the protopectins are enzymatically converted to pectinic acids. These compounds have greater solubility in water than the protopectins. Hence, the fruits become somewhat soft. In this way, the pectinic acids enzymatically convert to pectin. Solubility of pectin is still greater and the fruits become more soft. Lastly, the pectin molecules convert to pectic acid molecules by enzymes. Solubility of pectic acid in water is very high and hence, the fruits become very soft. This means that *their binding capacity with the adjoining cells goes very low* and this happens as methyl esters have been largely de-esterified. This increases the rate of water-solubility and the fruits become very soft in texture.

Among the different pectic compounds, pectin has a great role in health. Pectin does not act as a nutrient in the body but it is of much benefit in the cardiovascular and many other systems.

Most important function of pectin lies in that, it acts as one of the very important compounds in the group of *dietary fibres* and this has been discussed later. *In food industry, pectin is used in making jam, jelly, and marmalade.* When pectin solution is reacted with sugar, which is a dehydrating agent, a semi-solid substance known as *gel* is formed. This is called jelly. When the jelly contains fruit pulp, it is called jam. When the jelly or jam is prepared by mixing with fine pieces of citrus fruits (usually mandarin, *i.e.*, loose-skinned orange), the product is called marmalade. Pectin was first isolated by Henri Braconnot in 1825. Pectin is used in many other industries also, *e.g.*, adhesive, textile, lubrication *etc.* Fruits of some species are rich sources of pectin.

2.6. Digestion

Carbohydrates constitute 70 to 80 per cent of the diet for most people in the world. These contain all forms of carbohydrates, *i.e.*, monosaccharides, oligosaccharides and polysaccharides but monosaccharides are actually consumed in much lesser amount than the other two forms. Monosaccharides that are consumed by the human beings are mostly *glucose* and *fructose*, oligosaccharides that are mostly consumed come under disaccharides and these are *lactose, maltose* and *sucrose* while the largely consumed polysaccharides are *starch, dextrin, glycogen* and *cellulose*. Among the Indian people in particular, *starch forms the bulk of dietary carbohydrate* and *supplies 80 per cent of the calories.*

For utilization of carbohydrates in the body, it is essential that they should at first be *digested*. Only after digestion of the carbohydrates, the digested products would be absorbed in the body. The term, digestion (dig to be pronounced as digit) is actually derived from the Greek word, *digestio,* which means taking apart. This denotes breaking down (= cleavage), *i.e.*, catabolyzing (= hydrolyzing) into simpler forms, which are monosaccharides. The monosaccharides if any, taken through food, *e.g.*, glucose, fructose *etc.* are evidently not to be digested, *i.e.*, not required to be hydrolyzed and hence, they are directly absorbed. Thus, *it is only the oligosaccharides* and the *polysaccharides that are needed to be digested.*

It has earlier been stated under the heading on Starch (2.5.2 – C : polysaccharide – starch) that starch is the principal form of carbohydrate that the human beings consume from foods (= staple foods). Other forms of carbohydrates such as disaccharides that are taken are although relatively easily hydrolyzed, hydrolysis of starch is a complex phenomen. The process involves series of steps and are briefly outlined in the following:

First of all, it should be remembered that *digestion (= hydrolysis) of starch takes place by the enzyme,* which is known as *amylase.* The amylase enzyme is present in two parts of the body. These are, (i) in the *saliva* which is present in the *mouth* and (ii) in the *juice* which is secreted by the *pancreas.* Accordingly, they are respectively termed as *salivary enzyme* and *pancreative enzyme.* Salivary enzyme is more commonly termed as *ptyalin.*

Amylase enzymes which break down starch (= carbohydrate) molecules is actually a group of enzymes. That is, there are three forms of this enzyme, which are, *alpha-amylase, beta-amylase* and *gamma-amylase*. Alpha-amylase (Enzyme Commission Number 3.2.1.1) is also known by other alternative names which are, 1, 4–alpha-D–glucan–glucanohydrolase and glycogenase. This is a calcium *metallo-enzyme*, which means that calcium ions are necessary for its functioning and without calcium, it cannot function. It acts at random locations on the long chain of a starch molecule. It has been stated earlier under the heading, 2.5.2 that the starch molecule is actually a condensation product of two types of carbohydrates, which are, (i) amylose and (ii) amylopectin. The alpha-amylase hydrolyzes the *amylose* and on hydrolysis, gives rise to the carbohydrates, *viz., maltotriose* (= a trisaccharide) and *maltose* (= disaccharide). It hydrolyzes the *amylopectin* and gives rise to the carbohydrates, *viz., maltose, glucose* (= a hexose) and the "limit dextrin" (= a hexosan polysaccharide). This enzyme (alpha- amylase) is the major enzyme for digestion of carbohydrate and it functions at pH 6.7–7.0 (alpha–enzyme is present in plants and some fungi and bacteria).

The beta-amylase (EC No. 3.2.1.2) is also known by other names, *i.e.*, 1, 4–alpha-D–glucan malto–hydrose; glycogenase; saccharogen amylase *etc.* is actually synthesized by some species of fungi, bacteria and higher plants and not in the human beings. Starting from the non-reducing end, beta-amylase catalyzes hydrolysis of the second alpha- 1, 4 glycosidic bond. The gamma-amylase enzyme (E.C. No. 3.2.1.3), which is known by other names as glucan 1, 4–alpha glucosidase and amylo-glucosidase splits (breaks down) alpha (1–6) glycosidic linkages and the last alpha-(1– 4) glycosidic linkages of amylose and amylopectin at their non-reducing end and liberate glucose.

The process of digestion of carbohydrate in different organs is as follows:

(i) In the Mouth

The cavity of mouth is called *bucco-pharyngeal cavity*. In this cavity, the solid food is mechanically broken down by teeth (in the absence of teeth, gums do the function but less effectively) and more importantly, the mouth cavity secretes valuable substance which is called *saliva*. The saliva comes from three pairs of salivary glands, which are known as *parotids, submaxillary* and *sublingual glands* (Figure 3) and oral mucous gland. Saliva contains *plentiful alpha-amylase enzymes* and these are known as salivary amylase enzymes. These enzymes hydrolyze 1, 4– alpha linkage part of the starch but not the 1, 6–linkage part. On hydrolysis of the starch molecules by ptyalin, *i.e.*, salivary amylase, other carbohydrates, such as maltose, maltotriose, dextrin and a very small amount of glucose are liberated. The characteristic features of the ptyalin are, it acts on boiled starch only, its reaction takes place in slightly acidic condition, at a temperature of around 45°C and in presence of some salts, *e.g.*, chloride. It is stated that ptyalin contains very small amount of maltase enzymes but this is not confirmed.

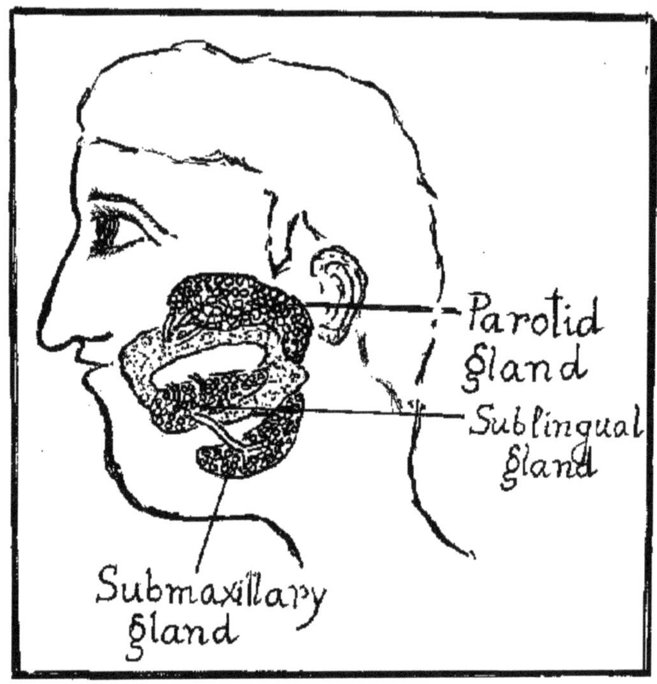

Figure 3. Salivary Glands.

The food which is chewed in the mouth becomes softened on mixing with the juicy saliva. This softened food is called *chyme*. The chyme therefore, gets plenty of salivary amylase enzymes and is able to hydrolyze, *i.e.*, digest the starch molecules. But unfortunately, the chyme does not remain in the mouth cavity for long time for which, very little part of the starch present in the food gets hydrolyzed. (Some children have the habit of chewing the food incompletely. They should be learned to chew thoroughly such that the food is mechanically ground, thoroughly mixed with saliva and remains in the mouth for some time).

(ii) In the Stomach

From the mouth cavity, the chyme (= softened food) passes on downwards to the stomach via the oesophagus pipe. The *stomach does not contain any amylase enzyme and hence, it cannot hydrolyze any starch molecule*. On the other hand, the stomach contains much of *hydrochloric acid* (HCl) and this is in appreciably high concentration. Hence, the salivary enzymes which remain mixed with the food (= chyme) is in fact, denatured, *i.e., inactivated by the hydrochloric acid of the stomach* and the enzymes are lost when the food comes to the stomach. However, it may be stated that the hydrochloric acid is not able to reach the *upper part* of the stomach. Accordingly, the chyme that comes downward upto the upper part of the stomach is not prevented against hydrolysis by the salivary enzyme. But this constitutes a very little part only. It may therefore, be commented upon that due to insalivation of food (mixed with saliva), some part of the starch is hydrolyzed. The starch that is hydrolyzed by the salivary amylase usually does not exceed 30 per cent while 70 per cent remains undigested.

Nevertheless, it may be stated that although the hydrochloric acid of the stomach inactivates the amylase enzymes present in the chyme, it renders at least one benefit of carbohydrate digestion. That is, this acid hydrolyzes (= breaks down) sucrose molecules, which is the common sugar. Thus, the molecules of common sugar if any reach the stomach, the hydrochloric acid will hydrolyze (= this is chemical hydrolysis and not enzymatic) that sugar, such that one molecule of sucrose will be broken down to liberate one molecule, each of glucose and fructose.

(iii) In the Small Intestine

From the stomach, the chyme (which is at first mixed with saliva and then with hydrochloric acid) is then passed on to the first part of the small intestine, *i.e.*, *duodenum* via pyloric orifice. In this organ, the *chyme* (this should not be regarded as chyme any more when it is mixed with hydrochloric acid in the stomach) receives a highly valuable juice from the pancreas (Figure 4). Pancreas is a spongy gland which measures about 15 cm in length and this is located behind the stomach in the upper part of abdomen. This is pear-shaped and consists of head, neck and tail. It has no connective tissue and is enclosed by a loose tissue which divides it into many lobules.

Interestingly, the pancreas gland functions *both as an exocrine gland, i.e., having ducts as well as an endocrine gland, i.e., having no duct (= ductless gland) in it*. Thus it is a mixed type of gland, *i.e.*, a dual purpose gland like that of liver.

The exocrine glands of the pancreas (= acini) are such that each one is provided with its own duct. Each duct anastomoses (= communicates between two parts, either directly or through channels) to form the main duct (= duct of Wirsung). This duct joins the common bile duct and opens at the duodenum (= hepato-pancreatic ampulla). Accessory pancreative duct (= duct of Santorini) also exists in some people and this duct opens into the duodenum directly.

The endocrine gland of the pancreas consists of *islets of Langerhans* (Figure 4). The islets may be 1 to 2 million in number and they consitute 1–2 per cent of the weight of the total weight of pancreas. Each islet is again composed of four kinds of cells. These are, alpha- (or α) cells (comprising 20 per cent), beta- (or β) cells (70 per cent), delta- (or δ) cells (5 per cent) and F cells (less that 5 per cent). These cells secrete specific *hormones, e.g.*, alpha-cells secrete *glucagon*, beta cells secrete *insulin*, delta-cells secrete *somatostatin* and F cells secrete *pancreatic polypeptide*. Among these hormones, insulin and glucagon are primarily concerned with regulation of glucose, somatostation is primarily concerned with paracrine regulation (insulin) and also secretion of glucagon while the function of the pancreatic polypeptide has not been clearly known. Glucagon raises blood glucose level, insulin lowers blood glucose level and somatostatin inhibits secretion of insulin, glucagon, growth hormone from the anterior pituitary and gastrin from the stomach.

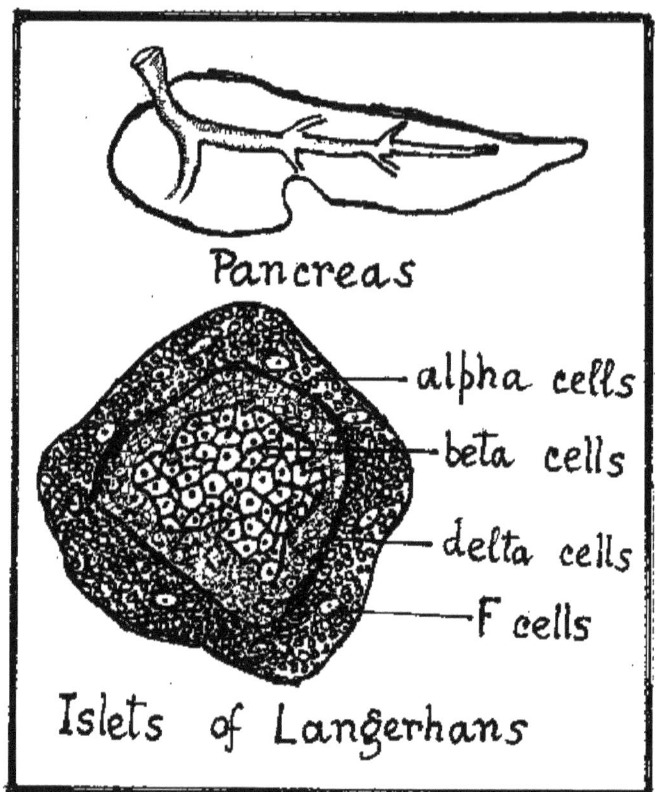

Figure 4. Pancreas and Cells in the Islets of Langerhans.

The *exocrine part of the pancreas gland is concerned with digestion of food.* That is, it secretes a valuable juice in the duodenum which is known as *pancreatic juice.* The pancreatic juice contains *varieties of enzymes and these enzymes are able to digest, i.e., hydrolyze complex organic compounds which are present in the food and thus, food is digested.* The juice, which is somewhat alkaline in reaction due to the presence of bicarbonate ions is secreted through papillae (in some persons, may be through accessory ducts as stated above). The secretion is regulated by the hormones, *secretin* and *cholecystokin* which are produced by the walls of duodenum.

Among other enzymes, the pancreatic juice contains *plenty of amylase enzymes,* known as pancreatic amylases, which are able to digest, *i.e.,* hydrolyze the starch molecules. Again, the food that comes down to the duodenum from the stomach contains plenty of starch molecules which remain as undigested (not hydrolyzed). This is because, only little portion of the starch molecules have actually been digested by the saliva (= salivary amylase) present in the mouth.

From the mouth, when the food comes down to the stomach, no part of the starch is digested as the stomach does not contain any amylase enzyme. Therefore, *major part of the starch molecules in the food remains undigested when it comes down to the duodenum from the stomach.* But the duodenum receives the amylase enzymes from the pancreatic juice. Accordingly, most part of the undigested starch which remains in the food is digested here on getting the pancreatic amylase enzymes. It may also be mentioned that the pancreatic amylase is highly efficient and *much more effective than the salivary amylase.* This enzyme is able to act both on boiled and unboiled starches while the salivary amylase acts on boiled starch only. The pancreatic enzyme is also not denatured easily by any agent. Its action is much more rapid than salivary amylase. For effective functioning of the pancreatic enzymes, pH should be 6.7–7.0, temperature should be around 45°C and chloride salts should also be present.

Pancreatic amylase hydrolyzes starch (= breaks down) to form alpha-dextrin, maltose and maltotriose. The hydrolytic products of alpha-amylase and alpha-dextrinase along with disaccharides to their corresponding monosaccharides by the enzymes, maltase, isomaltase, sucrase and lactase that are present in the *micro-villi* of the small intestine. (The *villus* is a small fold or hair-like projection of some mucous membrane and villi is the plural form. Micro-villi are vary small. A micro-villus is extremely small and is about 100 nm in diameter and 100–2000 nm long, A microvillus being very small can be seen through an electron microscope only. In ordinary light microscope, it collectively gives a fuzzy appearance on the surface of the epithelium. For this fuzzy appearance, collection of microvilli is termed as *brush border* as it looks like the bristles of a painting brush. Brush border cells are found in the small intestine, kidney and also large in intestine. Brush border increases surface area of cells). Micro-villi are very helpful as they increase the *surface area* and thereby, increase rate of absorption. It may be concluded that major part of the starch though may not be whole is hydrolyzed by the pancreatic enzyme and this takes place in the small intestine.

In respect of pancreatic secretion, an important point should be noted. That is, the pancreative juice not only secretes the amylase, *i.e.*, the starch-hydrolyzing enzyme but also contains other enzymes, which are, *trypsinogen, chymo-trypsinogen, elastase, carbopeptidase, nucleases* and *lipase.* These enzymes hydrolyze protein and fat molecules. Thus, hydrolysis of protein and fat is done along with hydrolysis of starch. Pancreatic enzyme has therefore, a great role in the body in hydrolysis, *i.e.*, digestion of many substances.

(iv) In the Large Intestine

Most of the starch is although hydrolyzed in the small intestine, *some portion may still be left* when the food comes down to the large intestine. But the large intestine has no mechanism to hydrolyze starch. In fact, it is concerned with absorption of water and vitamins, lowering of acidity, production of antibodies *etc.* Hence, the *resistant starch* which is usually obtained from some foods like potato, bean, oat, wheat flour *etc.* that remains undigested is difficult to be digested in the large intestine. However, it needs mentioning that plentiful bacteria which may be of 700 species inhabit in the large intestine. These bacteria which are known as "body-friendly bacteria" (= commensal bacteria) may catabolize (= hydrolyze) the undigested starch under anaerobic condition. Due to this process, which is known as *fermentation*, the undigested starch may be hydrolyzed, liberating gases like hydrogen, carbon dioxide and methane and also short-chain fatty acids like acetate, propionate, butyrate *etc.* It may be mentioned that the fatty acids are rapidly used up, *e.g.*, butyrate is used by cells in the colon and acetate is absorbed into the blood and from there, it is taken up by liver and muscular or other tissues. Propionate though gives rise to glucose in some animals but this is not common in human beings.

Lastly, it may be opined that digestion of starch is in fact, a complicated process which is controlled by a large number of factors. Individual and hereditary factors also play part. Age of the individual is again, a factor of starch hydrolysis. For example, in the new-born infants, *pancreatic amylase is not produced upto six month of age* and hence, the infants are not able to digest starch upto six months. Can it be mentioned here that the social and religious custom among the Bengali Hindu families, which is known as *annaprashana* directs that a new-born baby should be given rice (= *anna*) as God's *prasad* only when he or she becomes six months old and not before that as it is not possible to digest the carbohydrate of the rice before six months of age ? May be that the custom being strictly followed, it is linked with religion.

2.7. Absorption

The term, absorption is derived from the Latin word, *absorptio*, which indicates taking up process of liquids by solids or of gases by solids or liquids.

Carbohydrate molecules can be absorbed only when they are in simplest form, *i.e.*, monosaccharids. Hence, the bigger carbohydrate molecules are necessary to be broken down at first and digested into simplest forms, *i.e.*, into monosaccharides. However, some of the disaccharides which although are not in simplest forms are also absorbed directly and their absorption takes place by diffusion but only to a small extent.

Absorption of the monosaccharides occurs at the highest rate in the *jejunum* as blood supply is richest at this part. But adequate absorption also takes place in the duodenum and ileum. The villi and the micro-villi which are present along the walls of the intestine greatly increase the surface area and hence, rate of absorption is also made increased. Absorption takes place into the mucosal cells of the small intestine and then goes to circulation via *portal vein* (= portal vein conveys blood to the liver) straight to liver. It may be mentioned that lymphs also aborb monosaccharides but to a very small amount only. Sugar concentration in the blood whether high or low does not make any difference on the rate of absorption of monosaccharides or disaccharides.

In the small intestine, absorption takes place by three processes. These are, *simple diffusion*, *active transport* and *facilitated transport*. In the case of *simple diffusion*, there occurs an increase in the intestinal lumen (= space within intestine) and this increase is greater than the blood sugar (glucose) level. As a result, osmotic difference occurs between them from higher concentration of glucose in the lumen to the lower concentration of blood sugar and thereby, the glucose diffuses. In the case of *active transport*, diffusion continues to go on until the sugar (glucose) concentration in the lumen comes to the similar level with that of blood and after that, glucose is transported by active transport. The glucose is bound here to a *carrier protein*, which is present in the outer membrane of the intestinal wall. The carrier protein gets attached with two sodium ions. Being attached with glucose and sodium ions, the carrier protein moves to the cell and finally they are released to the cytoplasm. The sodium ions should be at the lower concentration in the cytoplasm and hence, these are needed to be expelled out of the cytoplasm into the blood plasma. This takes place in exchange of potassium ions through ATP enzyme pump. This hydrolyses ATP for exchange of sodium ions with those of potassium ions. In case of *facilitated transport*, fructose molecules are transported by a protein carrier and this does not require any energy. Thus, the process facilitates transport but this is however, a slow process.

Regarding rate of absorption of different monosaccharides, it could be said that their rate differs according to the type of monosaccharides. Thus, if the rate of absorption of glucose absorption is theoretically considered to score as say 100, it would be 110 for galactose and very less, *i.e.*, 39, 15 and 9, respectively for mannose, xylose and arabinose.

It should be noted that if glucose is produced in very high amount by digestion, or comes to the body from external source, *i.e.*, high intake of food rich in sugar, the excess amount is not absorbed in the body. The excess glucose molecules are in fact, stored as *reserve materials* for future use. Glucose molecules are stored in the form of *glycogen*. As stated under the heading, 2.5.2 – (Disaccharides), glycogen is a glucose–rich disaccharide. It is largely present in liver and muscles but also in brain, kidney, white blood cells and in uterus during pregnancy and glucose molecules are released from glycogen when there is necessity.

Absorption of some carbohydrates, *e.g.*, dietary fibres does not take place as these are not digested. These are expelled out of the body trough faeces. The disaccharide, lactose is also expelled out from the gut in case of people who have lactose intolerance disorder. (See Heading, Heptose-Oligosaccharide-Lactose)

2.8. Nutritive and Other Health Benefits

Quite a number of benefits are provided by the carbohydrates in the body. Some important ones are stated below:

(a) Providing Energy

It has been stated earlier that carbohydrates are the basic sources of providing energy in the body. One gram each of monosaccharides, disaccharides and starch respectively provides 3.74, 3.95 and 4.18 Kcal of energy in the body.

Energy is a *vital necessity* for functioning of all organs and tissues. Carbohydrates obtained from food are hydrolyzed and due to that, *glucose* molecules are released, which give rise to energy. Stomach and the small intestine *absorb glucose* molecules and *deliver* those to the blood stream. When they have come to the blood stream, they are used to release energy, or they may also be stored if the quantity is high and more than required. However, the body requires the hormone, *insulin* for making use of glucose molecules to release energy, or to store them.

Functions of the insulin which is released from the islets of Langerhans of the pancreas have been stated earlier under the heading of digestion (2.6). Insulin allows glucose molecules in the blood to arrive at the cells, so that the glucose molecules breakdown and provide energy.

(b) Benefits of Dietary Fibres

Importance of nutrition to keep up health and defend against illness have gained much prominence. In course of time, it has however, been realized by the health scientists that even if a person gets all the nutrients in sufficient amount, there are *other chemical substances* which must have to be taken by him or her through food in order to sustain health and get riddance of certain types of morbidness, especially those of physiological nature. The dietary fibres are the chemical compounds which aim at meeting such requirements.

The term, diet (Greek : *diaita*) refers to the solid or liquid forms of food that is consumed by a human being or any living organism. The major functions of diet are to provide *nutrients* in the body for healthy physical growth, mental development, getting energy for daily activities, protection or cure against illness, fulfilment of stomach, appeasement of satiety and evacuation. In the human beings, the diet may be *normal* which is habitually taken by an individual regularly or of *special type* when it is modified as desired for some particular purpose(s).

Fibre is a term (Latin : *fibra*) that refers to a thread, string or filament which is a thin, flexible material and the length of which is much greater than the width. Fibre is obtained from plants, *e.g.*, cotton, hemp *etc.*, animals, *e.g.*, silk, wool *etc.*, minerals like asbestos or synthetic materials, such as nylon or dacron.

Apparently, it would therefore, seem that dietary fibres are food of such type that are fibrous in appearance. However, this is incorrect in the science of dietetics. In dietetics, dietary fibres refer to some types of *carbohydrate compounds*, in other words, some *specific polysaccharides* that are obtained only from plant (botanical) source and not from any animal. In fact, dietary fibres is a collective term which signifies that it is a *mixture of some specific polysaccharides* which are the components of it. Except a few, these long chain glycans (= polysaccharides) are not digested or are digested to a very little extent by the enzymes secreted by the gastro-intestinal tract. Hence, these are mostly the *indigestible part of the carbohydrates*. The nutritional scientists have *not considered* the dietary fibres to include under the category of nutrients. This is because, these foods, although they are aliments do not provide any nutrition in the body like proteins, other form of carbohydrates, fats, vitamins and minerals and the energy if any released, is insignificant. But a section of nutritionists is not in favour of this argument and they opine that the dietary fibres should very much come under the category of nutrition for the enormous benefits rendered by them in the body. However, the controversy has not been settled and till now these compounds have not been textually recognized as nutrients.

Whatever may be the viewpoint, the term, dietary fibre is a *misnomer* and blatantly, a misleading term. Some people call these victuals as *roughage* or *roughage food* but this term also appears to be unfortunate. Some other people prefer to term the dietary fibres as *bulking food*. The reason is, they absorb much water in the gut to become much swelled and thereby, add bulk to them. The term is though not erring and has been on use erstwhile, could hardly be said to signify denotation.

(i) Components

There have been a variety of carbohydrates that are considered as dietary fibres and without hesitation, they are all polysaccharides. The common ones may be named as cellulose, hemicellulose, pectin, gum, mucilage, beta-glucan, lignin *etc.* At later time, some other carbohydrates, *e.g.*, dextrin, chitin *etc.* have also been included in the list by some. However, opinions vary according to dietitians and which glycans (= polysaccharides) should actually be recognized as dietary fibres remains controversial.

(ii) Types

In accordance with solubility in water and functions, dietary fibres are classified into two groups. These have been termed as *soluble dietary fibres* and *insoluble dietary fibres*. The differences between these two types of dietary fibres are as follows:

Soluble dietary fibres are soluble in water but they give a viscous appearance. They are *prebiotic*, which means that they encourage (promote) the growth of such species of bacteria which are health-friendly (= commensal) and are of great benefit to the digestive system of the person. By the action of these anaerobic bacteria, the soluble forms of dietary fibres are readily *fermented, i.e.*, the bacteria respire *anaerobically* taking the fibres as their food (= substrates). Due to such fermentation process which takes place particularly at the lower part of the gut, some energy is given off and *physiologically active end–products* are formed which are of much benefit. However, gas is also produced by them (= flatulence) which gives some discomfort. Some examples of soluble dietary fibres may be stated as pectin, gum, mucilage, beta-glucan and large proportion of hemicellulose.

Insoluble dietary fibres are on the other hand, not soluble in water. Although anaerobic bacteria act on them and use as substrates but they respire very slowly and hence, very slow fermentation takes place. Accordingly, the end-products that may be formed are very low in amount and they are not physiologically active.

Examples of the most common insoluble dietary fibres are cellulose, lignin and some protions of hemicellulose.

(iii) Physical and Chemical Properties

Dietary fibres have high *cation binding capacity.* Most of them are not digested except a few. Hydrolysis, *i.e.,* breakdown by enzymes may take place in them when their chemical structure is broken, *i.e.,* they are converted to short-chain carbohydrates like oligosaccharides and may also be monosaccharides. As these fibres are not much digested, they do not produce energy except to a small extent. However, as stated above, some energy is liberated by the soluble dietary fibres when anaerobic bacteria grow on them and respire, *i.e.,* when the soluble fibres are fermented. But the amount of energy that is produced by this action is very low and it may be about 2 calories only, i.e., 8.5 kilojoules due to fermentation of 1 gram of soluble fibre.

The dietary fibres make some change in the nature of the contents of the gastro-intestinal tract and restore such condition that other constituents, which are nutrients may get absorbed. Soluble dietary fibres reduce sugar accumulation in blood after their intake. They also tend to *normalize the lipid level* in the blood. Short-chain fatty acids are also produced by them as by-products and these by-products have wide range of physiological activities. *Lowering of blood sugar level* is also done by the soluble dietary fibres but the exact mechanism is not known.

(iv) Benefits

Both the soluble and the insoluble dietary fibres render so plentiful benefits in the body that the health scientists are now widely recommending food which contain these substances to each person every day, barring the new-born infants. The health scientists opine that their importance is *no less than the nutrients.*

In respect of soluble dietary fibres, first of all it may be stated that *they absorb large amount of water in the stomach and intestine and become viscous.* This causes delay in secretion from the gut and thus, reduces digestion. As a result of this, *absorption of glucose is delayed.* Due to delayed absorption of glucose, secretion of *insulin* hormone from the pancreas is also delayed. Thereby, the insulin level in the blood is eventually lowered. The Glycaemic Index (GI) is thus, much on the lower side. Therefore, the soluble dietary fibres regulate sugar level in the blood and are much helpful to the people afflicted with type-2 (insulin non-dependent) diabetes. Dietary fibres act as base on which beneficial bacteria grow in the stomach (prebiotic).

The soluble dietary fibres have a great role also in *preventing higher accumulation of cholesterol and particularly, the low-density lipid (= LDL) cholesterol* which is greatly harmful in cardiovascular health. By delaying digestion of food in the stomach, they indirectly cause delay in absorption of cholesterol. In fact, they *bind* the bile salts and prevent their reabsorption. This results in decrease of cholesterol level that circulates. The soluble fibres have also the properties to yield *short-chain fatty acids* which are formed due to *fermentation* caused by the bacteria in *colon*. The circulating biles are carried to the liver and the cholesterol synthesis is inhibited there or retarded. The soluble fibres remove bile and natural sterols from the body and excrete them through faeces. By adding *bulk* to the diet, they make the person feel full faster. It may also be mentioned that although the dietary fibres are said to provide no energy in the body, some energy although very low is given by them.

In regards to the insoluble dietary fibres, a great advantage of them is *relieving of constipation*. Chronic form of constipation is a serious health hazard and leads to many distressful sufferings notably, haemorrhoids, fistula and other proctological and rectal disorders. Many people afflicted with persistent constipation have the habit of taking laxative medicines indiscriminately. They should however, remember that *constipation corrected by drugs is constipation perpetuated*.

The present day opinion of the health experts is, although the exact cause of constipation is not clearly known till now but there are a number of factors which are associated with this abnormality. Above all, there is a strong support that regular consumption of insoluble dietary fibres is a *positive step* to alleviate irregular bowel habit which is chronic and stubborn. Chronic form of constipation may even be a cause of colo-rectal cancer. Toxins produced by the bacteria in the rectum may also bring about some types of cancerous growth and this may happen if the body is not cleared of the faeces. Therefore, irregular bowel habit, if occurs in a person for prolonged period need to be viewed with seriousness and as an organic way of treatment, intake of insoluble dietary fibres is a highly efficacious agent without causing side-effect.

Insoluble dietary fibres are of great help to prevent and cure the diseases, *diverticulosis* and *diverticulitis*. Diverticulitis is an inflammaion of one or more diverticulum (= sac or pouch on the wall of the intestine and especially in the colon), which causes stagnation of faeces in the small distended sacs of the colon with severe pain. The symptoms are less pronounced in diverticulosis. Intake of insoluble dietary fibres is of great benefit to prevent and heal the disorder. Among other benefits, maintaining health *of cardiovascular system* by regular consumption of insoluble dietary fibres is of special significance. Bariatric property, *i.e.*, prevention and control of obesity by both insoluble and soluble dietary fibres is also claimed by some of late.

(v) Requirements in the Body

The dietitians and health scientists have no hesitation that every person except the new-born babies must consume certain amount of soluble and insoluble dietary fibres every day.

According to the Institute of Medicine, United States Academy of Sciences, women and men of 50 years or less should take 25 and 38 gram of dietary fibre everyday and it should be 21 and 30 gram, respectively for persons who are more than 50 years of age. Under Indian condition, any definite recommendation is though not available, Gopalan *et al.* (2018) in their book, "Nutritive Value of Indian Foods," published by the National Institute of Nutrition, Hyderabad under the Indian Council of Medical Research have stated that every person should consume 25–40 gram of total dietary fibres, *i.e.*, a combination of soluble and insoluble dietary fibres, or 12–14 gram of total dietary fibre per 1000 kilo calorie energy.

Daily intake of soluble and insoluble dietary fibres by a person however, will depend upon a large number of factors. Some of the pertinent ones may be mentioned as age of the person, gender, occupation, costiveness if any in chronic form, existing gastro-intestinal disorder, obesity, hyper-or hypo-tension, hyper- or hypo-glycaemia, hyperlipidimia, cardiovascular, hepatic and renal functioning, haemoglobin level, colic pain, if any, pregnancy, lactation or other conditions. It should be remembered that whatever benefits of dietary fibres are there, excessive consumption particularly in unhealthy condition will make more harm than any good. Intake of insoluble and soluble fibres should also be *proportionate* and most of the dietitians are of the opinion that the ratio between these two should be 3 : 1. Lastly, it should not be forgotten that dietary fibres should not be taken paying less attention on food of high nutritional value.

(vi) Adverse Effect

Despite many benefits, some drawbacks of dietary fibres are also known. They tend to *bind some micro-nutrients* and thereby, prevent their absorption. When the soluble dietary fibres are used as substrate by the anaerobic bacteria, they are fermented and as a result, gas is produced. Too much accumulation of the gas causes flatulence (= abdominal gas) which gives discomfort and may exert pressure on the diaphragm and thus, on the heart and the lungs. Some people complain allergic reaction on taking dietary fibres. Suppression of appetite is brought about by some of the dietary fibres in some people.

(vii) Contents in different Food-stuffs

Dietary fibres are present only in food-stuffs that are of plant origin. *In animal source food, these are not found to be present.* In the victuals of plant-source, both soluble and insoluble dietary fibres are present but proportion of insoluble fibres remain at a higher level than the soluble fibres.

Gopalan, C., Rama Sastri, B.V. and Balasubramanian, S. C. (2018) in their book, "Nutritive Value of Indian Foods," published by the National Institute of Nutrition, Hyderabad, under the Indian Council of Medical Research, New Delhi and revised by Narasinga Rao, B. S. Deosthale, Y. G. and Pant, K. C. have presented the insoluble and soluble dietary fibres content in a variety of food-stuffs. The list has been represented in the Table 2.

Table 2. Insoluble and Soluble Dietary Fibre Content of some Food-stuffs under different Groups. (Figures are in gram per 100 gram of edible portion)

Food-stuff	*Insoluble Dietary Fibre*	*Soluble Dietary Fibre*
Cereal grains		
Bajra	9.1	2.2
Jowar	8.0	1.7
Maize (dry)	11.0	0.9
Ragi (marua)	9.9	1.6
Rice	3.2	0.9
Wheat	9.6	2.9
Pulses		
Bengal gram (whole)	25.2	3.1
Bengal gram (daal)	12.7	2.6
Bengal gram (daal)	7.6	4.1
Bengal gram (whole)	15.4	4.9
Green gram (whole)	14.7	2.0
Green gram (daal)	6.5	1.7
Lentil (masur, whole)	13.5	2.3
Lentil (daal)	8.3	2.0
Red gram (arhar, daal)	6.8	2.3
Red gram (whole)	19.8	2.8
Leafy vegetables		
Agathi (bak phool)	6.3	2.1
Amarantha (kanta note)	3.1	0.9
Cabbage	2.0	0.8
Colocasia (leaves)	5.1	1.5
Coriander (leaves)	3.0	1.3
Curry leaves	13.4	2.9
Drumstick (leaves)	6.8	2.2
Fenugreek (leaves)	3.2	1.5
Mayalu (pui sak)	1.6	0.9
Paruppu (puruni sak)	2.9	1.0
Ponaganni (saranti sak)	6.9	1.0
Spinach	1.8	0.7
Tamarind (leaves)	9.4	1.2
Roots and tubers		
Beet root	2.6	0.9
Carrot	3.0	1.4
Potato	1.1	0.6

Food-stuff	*Insoluble Dietary Fibre*	*Soluble Dietary Fibre*
Radish	1.8	0.5
Sweet Potato	2.6	1.3
Yam (ghet kachu)	3.2	1.0
Colocasia (stem)	2.3	0.7
Other vegetables		
Banana (stem, thor)	2.0	0.2
Bitter gourd	3.2	1.1
Bottle gourd	1.7	0.3
Brinjal	4.6	1.7
Broad bean	6.7	2.1
Couliflower	2.6	1.1
Cluster bean (guar)	4.2	1.5
Cucumber	2.0	0.6
Drumstick (pod)	4.8	1.0
Giant chilli	2.0	0.2
Okra	2.6	1.0
Other vegetables		
Onion stalk	3.7	1.4
Pea (green)	7.2	1.4
Plantain	2.6	0.9
Ridged gourd	1.4	0.5
Snake gourd	1.6	0.5
Tomato	1.2	0.5
Nuts and oilseeds		
Soybean	17.9	5.1
Coconut (fresh)	12.7	0.9
Gingelly seed (til)	13.6	3.2
Groundnut	8.5	2.5
Mustard	10.2	3.4
Condiments and spices		
Aniseed	34.3	9.1
Cinnamom (bark)	44.6	3.9
Onion (bulb)	1.7	0.8
Cardamom	20.4	2.6
Clove	28.9	6.2
Coriander	42.5	4.9
Cumin (jira)	25.2	4.8

Food-stuff	*Insoluble Dietary Fibre*	*Soluble Dietary Fibre*
Fenugreek	28.6	20.0
Garlic (bulb)	2.6	2.5
Black pepper	32.4	2.9
Opium poppy (Posto) seed	22.4	11.4
Turmeric	17.6	2.4
Fruits		
Papaya (ripe)	1.3	1.3
Ziziphus (ber, bor, kul)	2.8	1.0
Emblica (aonla, amlaki)	5.8	1.5
Apple	2.3	0.9
Banana (ripe)	1.1	0.7
Cherry (red)	0.9	0.6
Date (dry)	6.9	1.4
Fig	2.6	2.4
Grape (green variety)	0.8	0.4
Guava	7.1	1.4
Jackfruit (ripe)	2.1	1.4
Black plum	2.6	0.9
Sweet lime juice	1.3	1.4
Mango (ripe)	1.0	1.0
Musk melon	0.5	0.3
Water melon	0.3	0.3
Orange (mandarin)	0.6	0.5
Peach	1.1	0.5
Pineapple	2.3	0.5
Plum	1.7	1.1
Sapota	9.1	1.8
Custard apple	4.0	1.5
Strawberry	1.6	0.7

(viii) Crude Fibre

Dietary fibres should not be considered to be same as *crude fibres*. It should be noted that crude fibres are in fact, different from dietary fibres. The term, crude fibres were used earlier but not now although still on use in some food laboratories. These are referred to the fibres (food) which are not soluble with acid and alkali. That is, crude fibres constitute the residue, which is left after laboratory procedure of digesting (= heating) the food-stuffs sequentially with 0.255 *Normal* sulphuric acid (H_2SO_4) and 0.313 *Normal* sodium hydroxide (NaOH) solution, followed by drying the sample at 104°C overnight and igniting at 600°C for 3 hours. By this process, the compounds, *viz.*, sugars, starch, protein, lipid and some part of lignin and carbohydrates are removed. The highly indigestible part which is left as residue is referred to as crude fibre and this is composed of cellulose, hemicellulose, lignin, ash and tannin.

The crude fibres, which consist of mixture of *insoluble fibres have in fact, not of much health benefit.* In most cases, they do not contain more than half to seventh part of the dietary fibres. At the present time, dietary fibres are given importance in food science and dietetics with little emphasis on crude fibres.

(c) Cardiovascular Benefits

Many people are afraid of carbohydrate-rich foods for the belief that such foods are harmful to cardiovascular functioning and also bring about obesity in the body. They should take note of the fact that carbohydrate is like a coin which has two sides. That is to say, some carbohydrates are there, which give rise to relatively high energy in the body and if the energy is not expended by physical activity, extra fat may be added in the body to give rise to obesity or atherosclerosis. On the other hand, there are polysaccharides and more so, the complex polysaccharides which are recognized as dietary fibres. Carbohydrates of this class protect the body against cardiovascular malfunctioning and weight gain.

(d) Improving Muscle Mass

Intake of adequate and right type of carbohydrates is of great benefit to persons having ill-developed muscles. This is also of great significance for athletes and to those who are to undertake heavy physical work.

(e) Functioning of Brain

There is experimental evidence that *among other nutrients, glucose has a vital role* for improved functioning of the brain and nerves. This hexose sugar is the primary fuel for the muscle and brain and has been stated to be "crucial to the growth and development of the infants and young children". Fats cannot provide fuel to the brain as the neurons cannot burn fat. Persons who regularly consume sub-optimal amount of carbohydrates in their diet may develop *brain fog* (= feeling that thinking, understanding and remembering processes are not as desired) and other nervous complications.

(f) Serotonin Production

Serotonin (5 – hydroxytryptomine) is produced by some types of carbohydrates. It exists in gastrointestinal tract (mucosa), mast cells, blood platelets, brain and other parts. (It is also present in cancerous cells). Although serotonin acts as a vasoconstrictor, which has adverse effect in vascular organs, it is a *neurotransmitter* and has many good effects. It improves *mood* of a person, reduces anxiety, regulates appetite, digestion, sex desire, sleep-wake cycle *etc.*

(g) Improved Metabolism

Carbohydrates support normal *thyroid hormones* (thyroxine, tri-iodo thyroxine *etc.*) and overall metabolic functions, resulting more expenditure of energy.

(h) Gustation

Hardly any individual could be found in the world who does not like sweet food. Sucrose (common sugar) is a carbohydrate that imparts sweetness in a food. Some other carbohydrates are also sweetening substanes as stated in Table 1, under the heading, 2.4. Sweetening agents are suitably blended with food preparations to dulcify (= sweeten) them.

(i) Less Risk of Cancer

Some of the carbohydrates are known to prevent certain types of cancers. This is of significance in carbohydrates of such types that have abundant *anti-oxidant properties*.

2.9. Requirement in the Body

It should be remembered that the carbohydrates belonging to the groups of monosaccharides and oligosaccharides are *primarily involved to yield energy* in the body. Therefore, how much carbohydrates of these two types should be consumed would depend on requirement and expenditure of energy of an individual. In many parts of India, it is noticeable that the poor people meet up 80 per cent of their energy requirement from *carbohydrate-rich food*. Consumption of rice (= refined carbohydrates) is also relatively high in many parts of south India. Too much consumption of carbohydrates is in fact, *not desirable* because such type of dietary habit tends to make sub-optimal consumption of other food which are rich in other nutritive components like protein, fat, vitamins and minerals. On the other hand, sub-optimal consumption of carbohydrates by a person particularly the monosaccharides and oligosaccharides may seriously act upto health and more so, if energy expenditure of him or her is not commensurate to the physical activity undertaken.

Any exact recommendation regarding daily intake of carbohydrates by a person is in fact, difficult to prescribe. Under Indian condition, the dietitians opine that the standard rule should be such that *a person should meet upto 10 per cent of calorie requirement from protein, about 25 per cent of calorie requirement from fat and the rest 65 per cent of calorie should come from carbohydrates*. However, the National Institute of Nutrition (under the Indian Council of Medical Research) in Hyderabad provided the following recommendation in 1989.

2.10. Adverse Effects

Intake of carbohydrates may adversely act upon health if they are not taken by a person according to his or her requirements. It is often said by some people that intake of carbohydrates leads to obesity. But this is *only a fallacy*. As a matter

Table 3. Daily Requirement of Energy (in terms of Kcal/d) for the Indian People

Group	*Particulars*	*Body Weight (kg)*	*Net Energy Required (Kcal/d)*
Men	Sedentary work	60	2425
	Moderate work	60	2875
	Heavy work	60	3800
Women	Sedentary work	50	1875
	Moderate work	50	2225
	Heavy work	50	2925
	Pregnant Woman	50	+ 300
	Lactating Woman	50	+ 300
Infants	0 – 6 months	50	+ 550
	6 – 12 months	50	+ 400
	0 – 6 months	5	108 per kg
	6 – 12 months	8.6	98 per kg
Children	1 – 3 years	12.2	1240
	4 – 6 years	19.0	1690
	7 – 9 years	26.9	1950
Boys	10 – 12 years	35.4	2190
Girls	10 – 12 years	31.5	1970
Boys	13 – 15 years	47.8	2450
Girls	13 – 15 years	46.7	2060
Boys	16 – 18 years	57.1	2640
Girls	16 – 18 years	49.9	2060

of fact, monosaccharides and oligosaccharides when taken would hydrolyze in the body to yield ultimately glucose molecules. Glucose is energy-giving hexose sugar and it oxidizes in the cells to liberate energy. It has been stated earlier that, if the energy gained is not adequately expended by physical activity, it will lead to accumulation of fat in the body, resulting weight gain or obesity but for this, why carbohydrates be blamed ? Accumulation of high amount of glucose in blood plasma may result if insulin production in a person is not enough to burn the sugar. For this, the pancreas should be blamed rather than carbohydrates. Many health experts suggest that regular intake of high amount of carbohydrates produce metabolic syndrome X or insulin resistance syndrome. Atherosclerosis, breast cancer in women, prostate cancer, hyperlipidmia and some other disorders have also been said to be associated with high carbohydrate consumption. However, such opinions seek strong experimental support for confirmation.

2.11. Carbohydrate Content of some Food-stuffs

Carbohydrate content of some food-stuffs as listed by Gopalan, C., Rama Sastri, B.V. and Balasubramanian, S.C. in their book titled, "Nutritive Value of Indian Foods," 2018 (Revised by Narasinga Rao, B.S., Deosthale, Y.G. and Pant, K.C.) and published by the National Institute of Nutrition, Hyderabad, under the Indian Council of Medical Research has been presented in the following: Carbohydrate content of some selected aliments have only been presented. For the other food-stuffs listed, the original title may be referred to.

Table 4. Carbohydrate Content of some Food-stuffs under different Groups (Figures are in gram per 100 gram of edible portion).

Cereal grains			
Bajra	67.5	Barley	69.5
Italian millet	60.9	Jowar	72.6
Maize (dry)	66.2	China dhan	70.4
Ragi (marua)	72.0	Rice (parboiled, handpounded)	77.4
Rice (parboiled, milled)	79.9	Rice (raw, hand pounded)	76.7
Rice (raw, milled)	78.2	Rice (bran)	48.4
Rice (flakes, chira)	77.3	Rice (puffed, muri)	73.6
Samai	65.5	Wheat (bulgar, parboiled)	77.2
Wheat (whole)	71.2	Wheat (flour)	69.4
Wheat (flour, refined, maida)	73.9	Wheat (semolina, suji)	74.8
Wheat (bread, white)	51.9		
Pulses			
Bengal gram (whole)	60.9	Bengal gram (daal)	59.8
Black gram (daal)	59.6	Cowpea	54.5
Field bean (dry)	60.1	Reen gram (whole)	56.7
Grean gram (daal)	59.9	Horse gram (whole)	57.2
Khesari (daal)	56.6	Lentil (masur)	59.0
Moth beans	56.5	Pea (green)	15.9
Pea (dry)	56.5	Rajmah	60.6
Red gram (daal)	57.6	Soybean	20.9
Leafy vegetables			
Amaranth	2.0	Bathua	2.9
Beet green	6.9	Brussel's sprout	7.1
Cabbage	4.6	Colocasia (leaves)	8.1
Curry (leaves)	8.7	Drumstick (leaves)	12.5
Fenugreek (leaves)	6.0	Lettuce	2.5
Parsley	13.5	Spinach	2.9
Marsilea (susuni)	4.6		

Roots and tubers			
Arrow root (flour)	83.1	Beet root	8.8
Carrot	10.6	Colocasia	21.1
Dioscoria	18.1	Onion (big)	11.1
Onion (small)	12.6	Radish (pink)	6.8
Sweet potato	28.2	Topioca	38.1
Yam (ghet kochu)	26.2		
Other vegetables			
Ash gourd	1.9	Bitter gourd (small)	10.6
Bitter gourd (large)	4.2	Brinjal	4.0
Cauliflower	4.0	Drumstick (pod)	3.7
Drumstick (flower)	7.1	Giant chilli	4.3
Jackfruit (unripe)	9.4	Knol khol	3.8
Okra	6.4	Leek	17.2
Papya (unripe)	5.7	Plantain (fruit)	14.0
Plantain (flower, mocha)	5.1	Plaintain (stem, thor)	9.7
Pumpkin	4.6	Ridged gourd	3.4
Snake gourd	3.3	Tinda	3.4
Tomato	3.6		
Nuts and oilseeds			
Almond	10.5	Cashew (kernel)	22.3
Coconut (fresh)	13.0	Gingelly seed	25.0
Groundnut	26.1	Mustard (seed)	23.8
Pistachio (pesta kernel)	16.2	Walnut	11.0
Condiments and spices			
Asafoetida	67.8	Cardamom	42.1
Cloves	46.0	Coriander	21.6
Cumin (jira)	36.6	Garlic	29.8
Ginger (fresh)	12.3	Mace	47.8
Nutmeg (fruit)	28.5	Jaiphal (rind)	11.2
Jowan (omum)	24.6	Poppy seed (posto)	36.6
Turmeric	69.4		
Fruits			
Emblica (aonla)	4.5	Apple	13.4
Avocado	0.8	Wood apple (bael)	31.8
Banana (ripe)	27.2	Bullock's heart	15.7
Cashew apple	12.3	Cherry (red)	13.8
Grape (blue)	13.1	Guava	11.2
Jackfruit (ripe)	19.8	Black plum (jamun)	14.0

Litchi	13.6	Musk melon	3.5
Water melon	3.3	Orange (mandarin)	1.9
Papaya (ripe)	7.2	Pear	11.9
Pineapple	10.8	Plum	11.1
Pomegranate	14.5	Raisin	74.6
Sapota	21.4	Custard apple	23.5
Strawberry	9.8	Elephant apple (kath bael)	18.1
Ziziphus (ber, bor, kul)	17.0		
Fishes and sea food			
Bata (small)	2.2	Beley	2.9
Bhekti	3.0	Boaal	7.6
Chital	3.1	Crab muscle	3.3
Herring	2.2	Hilsa	2.9
Kalabasu (kalbos)	2.0	Katla	2.9
Khoyra	5.2	Lata	3.4
Mangri (magur)	4.2	Mahasol	1.0
Pabda	4.6	Parsey	4.3
Pomfret (white)	1.8	Prawn	0.8
Rohu (rui)	4.4	Shark	0.8
Singhi	6.9	Soal	2.2
Meat and poultry			
Beef muscle	0.2	Duck meat	0.1
Duck egg	0.8	Liver sheep	1.3
Snail (small)	3.7	Venison	1.9
Milk and milk products			
Ass's milk	6.5	Buffalo's milk	5.0
Cow's milk	4.4	Goat's milk	4.6
Human milk	7.4	Curd (cow's milk)	3.0
Chhana (from cow's milk)	1.2	Cheese	6.3
Sugars			
Sugarcane	99.4	Honey	79.5
Jaggery (gur from sugarcane juice)	95.0	Jaggery (gur from date plam juice)	86.1
Jaggery (gur from fan palm juice)	98.5	Jaggery (gur from coconut palm juice)	83.5
Beverage (Non-alcoholic)			
Neera	10.9	Sugarcane juice	9.1
Beverage (Alcoholic)			
Pachwai (of Assam)	5.8	Fermented palmyra palm juice (tari)	1.8

Chapter 3

Proteins

In the science of nutrition, protein ranks as the second important nutrient after carbohydrate. It provides structure to the tissues and muscles and also makes them repaired when necessary. Functionally, this nitrogen containing organic compound has indispensable role in the syntheses of enzymes, hormones and antibodies and in storage and transportation of molecules within cells on to other parts.

Whether protein should be regarded as the first nutritive component or it should be ranked as the second after carbohydrate remains to be a debated issue among the nutritional scientists. To settle the dispute, it may be commented upon that each and every nutrient has equal importance even if a nuance of it is required in the body. This is evident from the fact that complete absence of any nutrient stands in the way in healthy functioning of the body systems and thereby, effectuate malnutritional disorders. Nevertheless, it comes up to the mind of a biochemist and also to the nutritionist that, carbohydrate is the *base material* for the synthesis of other nutrients in the body. Hence, this compound seeks to receive higher rank than protein despite the fact that the term, protein has derived from the Greek word, *protos*, the English meaning of which is first. It is now known that this valuable nitrogenous compound is not only a nutrient but has a great role to defend and cure against many ailments in the body also and hence, it is regarded as a *nutraceutical* (= nutritive + pharmaceutical) substance.

3.1. Definition

The term, protein was given by Dr. Mulder in 1838 from the Greek word, *protos*, which means first. He proposed this name for this compound owing to the fact that it renders plentiful benefits in the body and hence, according to him, it is for *first requirement in the body* and thereby, the compound should justifiably bear the name, first. It was known that protein is not only of one type and there are various types of protein and all of them are of great help in the body.

Proteins are basically carbohydrate compounds, *i.e.*, their molecules contain the chemical elements, which are carbon, hydrogen and oxygen. But apart from these three elements, *there must be another element, which is nitrogen*. Hence, a protein molecule is such, in which, the elements, carbon, hydrogen, oxygen as well as nitrogen must be present. Thus, a protein molecule may be called a *nitrogen containing carbohydrate molecule*. However, not only carbon, hydrogen, oxygen and nitrogen but some other elements also remain present in protein molecules of some types but not in all types of proteins. If the other elements are present, their proportion is very small. In general, in a protein molecule, carbon may be present from 50 to 54 per cent, hydrogen 6 to 8 per cent, oxygen 20 to 25 per cent and nitrogen 14 to 20 per cent. Other elements if present, are usually less than 2 per cent.

The elements, carbon, hydrogen and oxygen along with nitrogen constitute some special types of acids which are known as *amino acids*. The nitrogen gives rise to ammonia on reaction with hydrogen and the ammonia reacts with carbon, hydrogen and oxygen to produce amino acids. The amino acids that are formed may combine with other elements.

Amino acids are of various types. They join together one after the other in chain by the process, known as *peptide bonds*. In this way, when many amino acids are joined together, one after the other by a separate peptide bond, the compound that is formed is called *protein*. However, when only many amino acids are joined and which may be 100 or more, it will be called a protein molecule. If the number of amino acids that are joined are relatively less, the compound should theoretically not be called protein.

3.2. Structure

A protein molecule is formed by joining (= combination, condensation, linkage) of many amino acids. Hence, each individual amino acid is called a *monomer* and the protein molecule which is formed by joining of all these monomers is called a *polymer*. Thus, a protein molecule is a polymer formed by amino acid monomers. Monomers are also called units or residues.

Joining of Amino Acids by Peptide Bonds

A peptide bond joins two amino acids. This is done in such a way that the alpha-carboxyl group (COOH) of one amino acid is joined (= condensed) with the alpha-amino group (NH_2) of another amino acid. In this joining process, one molecule of water (H_2O) is removed out. This is shown in Figure 7.

When only two amino acids are joined, the joined (condensed) product is called *dipeptide* (di = two), when three amino acids are joined, it is called a *tripeptide* (tri = three) and when many amino acids are joined, it is called a *polypeptide, peptone*

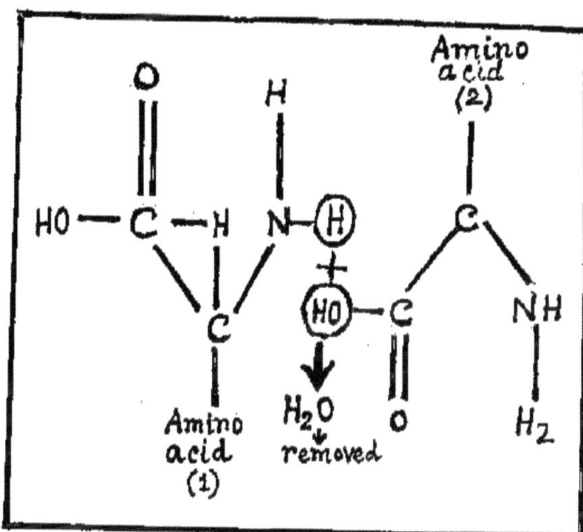

Figure 7. One Amino Acid Molecule is Joined with Another Amino Acid Molecule and One Molecule of Water is given off.

or *proteose*. In this way, when 100 or more amino acids are joined together, the condensation product of the amino acids should be termed a protein. Thus, a protein molecule (= polymer) is a chain of plentiful amino acid molecules, *i.e.*, units or residues (monomers). Accordingly, a protein is a highly nitrogenous carbohydrate complex compound.

In this context, it needs mentioning that the structure of the protein chain may be *primary*, *secondary*, *tertiary* or *quaternary*. The peptide bond and the amino acids in sequence gives rise to its primary structure. That is, the number of amino acid residues and their sequence determines its primary structure. The secondary structure of the protein chain is related to the spatial configuration of it. The chains of the protein molecules remain folded in the secondary structure. When the protein chains form coiled structure such that it becomes roundish in appearance, it gives tertiary shape of the protein molecule while in protein molecules having more than one chain, the inter-relationships between the consecutive chains determines its quaternary structure and the protein of the *tobacco mosaic virus* (TMV) which has 158 amino acid residues may be cited as an example.

Thus, it may be said that the compound, *protein is like a brick wall*. As a brick wall is constructed by joining, *i.e.*, cementing together the individual bricks on placing them one after the other, the protein molecule is also formed by joining the individual amino acid molecules one after the other. As in a brick wall, the bricks are strongly attached with the help of cement and sand mixture, so also in the protein molecule, the amino acids are strongly held, *i.e.*, joined together by means of *peptide bonds*.

3.3. Functions

Proteins play a great role in regulating various physiological processes in the body. Some examples may be stated as follows: The *plasma-proteins* maintain acid-base balance and osmotic pressure of fluids in the body. The *transport-proteins* in blood carry ingredients to various parts in the body. The haemoglobins in the red blood cells take up oxygen from lungs and deliver to the cells in different parts of the body and pick up carbon dioxide from the cells. Immunoglobins protect the body from the attack of harmful microbes. Strength to body parts in provided by the *collagen and the fibrin proteins*. Fibrinogen acts in coagulation of blood and thus, protects loss of blood in case of injury.

Proteins make good the loss sustained due to daily wear and tear of body tissues. They have essential role to provide fresh tissue for the growth of children and convalescents. For synthesis of enzymes, hormones and antibodies, the basic material is protein.

3.4. Chemical Properties

There are plentiful types of proteins and each one of them differs from the other in one or more physical or chemical characteristics. As compared to many other organic compounds, molecules of proteins are large or very large and hence, they are called *macromolecules* (macro = big). Molecular weight of proteins is also very high and ranges from 5000 in the *insulin*, which is the smallest known molecular weight among all proteins to 17,400 in milk protein, 63,000 in *haemoglobin* of human blood and is 60,000 in *tobacco mosaic virus* (TMV) which causes disease in tobacco and other plants. Because of so high molecular weight, most of the protein molecules *cannot diffuse (penetrate) through semi-permeable membrane*. As regards solubility, they are soluble in water, weak solution of salts, in dilute acids and dilute alkalis. Due to very large size of the protein molecules, they make *colloidal solution*, although some proteins are crystalline.

Since the protein molecules posses free electrical charged group, they can *migrate in an electrical field*. In containing both *amino* (NH_2) and *carboxyl* (COOH) groups, the proteins behave either as acid or as base, *i.e.*, they are *amphoteric* in nature. Coagulation on heating is very common for most of the proteins. As stated earlier, a protein molecule on hydrolysis, liberates the intermediate peptides and ultimately *those amino acids, i.e.*, with which it is formed. Proteins are able to make change of the colour of some specific reagents when reacted with them and such reagents are specially prepared and known as Biuret reagent, Millon's reagent *etc.*

Chemical *denaturation* is a property of the protein molecules. This means that by the action of some physical or chemical agents, the naturally occurring protein (= native proteins) molecules are disorganized and for this, the actual configuration, *i.e.*, the spatial arrangement of them is changed. Accordingly, their physical and chemical properties are also changed. The agents that cause such degeneration are heat, exposure to X-ray, ultraviolet ray, giving high pressure, vigorous shaking when in solution, addition of acid, alkali, enzyme *etc*. Denaturation causes either small or large change in the structure of the protein molecules. If the change is small, the denatured protein molecules may come back to its original structure and this is called *regeneration* or *renaturation* but when large change has taken place in the structure of the protein molecules, it cannot be brought back to the original structure.

Another property of the protein molecules is their *precipitation*, which takes place by the action of acids, alkalis, enzymes, alcohols, salts of some heavy metals such as lead, mercury, silver *etc*. in alkaline medium, some alkaloids *etc*. However, all proteins are not precipitated by the same agent, *i.e.*, the agents differ according to proteins in respect of precipitation.

3.5. Classification

Proteins are of plentiful types and to distinguish them from one form to the other, their grouping, *i.e.*, classification has its necessity. Classification of proteins is done in a number of ways but it may be commented upon that fully satisfactory system of classification is perhaps lacking as knowledge of their structure is not adequate. The conventional systems of classification have however, been presented below in brief.

In accordance with chemical structure, proteins are sometimes classified into two broad groups, which are termed as *fibrous proteins* and *spherical proteins*. In the fibrous proteins, the polypeptide chains are arranged in *sheets*, *i.e.*, strands while in the spherical proteins, the chains remain in folded condition and thereby, they have a *spherical* (= globular) shape. The fibrous proteins provide strength to the organs and collagen and keratin are common examples. The globular proteins on folding back the chains impart compact appearance and in maintaining such shape, they carry out many functions. Haemoglobin and immunoglobin are common examples of globular proteins. It has been seen in X-ray analysis that haemoglobin molecule has nearly spherical appearance with a diameter 5–6 nanometer.

The classification made on recommendation by the American Physiological Society, American Society of Biological Chemists and the British School on the basis of solubility properties and other physical and chemical differences is however, widely accepted now. In this system of classification, all proteins have been brought under three main groups which are known as *simple proteins*, *conjugated proteins* and *derived proteins*. Some of the physical and chemical properties and examples of these three groups have been stated as follows:

(a) Simple Proteins

These proteins are composed of only amino acid molecules and nothing else. Hence, these are also termed as *pure proteins*. Thus, when such a protein molecule is hydrolyzed (= broken down by enzymes), it liberates the intermediate peptides and ultimately those amino acid molecules which have constructed it. Some examples of pure proteins are given below:

(i) *Albumins* – These are soluble in water and salt solutions of neutral pH and are easily coagulated by acids, alkalis or heat. Some examples are egg albumin (= egg white), milk protein (= lacto-albumin), serum albumin, myoalbumin (in muscles), beta-amylase (in barley) *etc.*

(ii) *Globulins* – These are also soluble in neutral salt solution, insoluble in water, coagulated by heat or acid and are precipitated by solution of magnesium sulphate or sodium chloride. Some examples are ovak globulin (= egg-yolk), myosin (muscle protein), fibrinogen (= plasma protein), crystallin, serum globulin *etc.*

(iii) *Glutelins* – These are soluble in highly dilute acids and alkalis and insoluble in solvents of neutral pH. Glutenin of wheat is an example.

(iv) *Prolamines* – These are soluble in water, weak acids or alkalis or 50–70 per cent alcohol but not in absolute alcohol (= ethanol). Some examples are zein in maize, gliadin in wheat, hordein in barley, papain in papaya latex (= milky white exudation).

(v) *Histones* – For basic reaction, histone is commonly termed as *basic protein*. Histone is a constituent of chromosome. It occurs in haemoglobin and thymus gland also. It is soluble in water and dilute acids and is precipitated by ammonium hydroxide.

(vi) *Protamines* – Proteins of this class are strongly basic in reaction, have low molecular weight and soluble in ammonium hydroxide. In sperms and spawns of many fishes, protamines are found.

(vii) *Scleroproteins* – Proteins of this class have importance is giving structural strength to body parts, *e.g.*, keratin in nails and hairs, collagen of bones, ossein in teeth and bones *etc.* Solubility of them in any solvent is very low.

(b) Congugated Proteins

Proteins belonging to this class contain *some other components* (conjugation = joining together) along with amino acids and those non-amino acid components are called *prosthetic group.* There are many proteins which are conjugated and some examples are stated below:

(i) *Nucleoproteins* – These proteins are combined (= conjugated) with *nucleic acids.* Nucleic acids are of two types which are, ribonucleic acid (RNA) and deoxyribonucleic acid (DNA) and each one is composed of nucleotides. Each nucleotide has three parts, which are, a *nitrogenous base,* which is a purine or a pyrimidine compound, a *ribose (pentose) sugar* and a *phosphoric acid.* (These have been described later under the heading, 3.7).

(ii) *Chromoproteins* – The prosthetic groups of these proteins are some type of *colouring matter* (chrome = colour), *i.e.*, pigments. Examples are haemoglobin which contains the iron containing pigment, haem, rhodopsin in the retina of eye which contains the carotenoid pigment, flavoprotein containing riboflavin, cytochrome containing iron as the prosthetic group and the like.

(iii) *Glycoproteins* – These proteins have carbohydrates (glyco = carbohydrate) as prosthetic group. But the carbohydrate content is less than 4 per cent. Glycoproteins may be of two types, which are mucins (found is mucus, saliva, bile, skin, tendon, cartilage *etc.*) and mucoid (similar to mucin but differ in solubility and coagulability).

(iv) *Mucoproteins* – Similar to glycoproteins but carbohydrate content in them is higher and is more than 4 per cent. It is a complex of protein and muco-polysaccharide. Polysaccharide mostly contains hexose-amine.

(v) *Phosphoproteins* – Phosphoric acid is the prosthetic group in these proteins and hence, these are acidic in reaction. Casein in milk, vitellin in egg-yolk are some example. (Formerly called, nucleo-albumin).

(vi) *Lipoproteins* – Lipids (= fatty substance) are the prosthetic group of these proteins and occur in milk, egg, blood plasma *etc*. These are classified as very low density, low density, intermediate density and high density.

(vii) *Metalloproteins* – Some chemical elements derived from metals, *e.g.*, iron, copper, manganese, cobalt, magnesium *etc.* are combined (bound) as prosthetic group in these proteins. Many enzymes come under this group.

(c) Derived Proteins

The intermediate (= peptides) compounds that are obtained due to enzymatic break down (= hydrolysis) of protein molecules are known as derived proteins. Thus, derived proteins are hydrolytic products, *i.e., hydrolysates of proteins*. Hence, they do not occur naturally and are obtained only when proteins are hydrolyzed, *i.e.*, degraded. Derived proteins are sometimes made into the two groups, which are *primary* derived proteins and *secondary* derived proteins. Primary proteins are those in which, hydrolysis has made only slight changes, proteins, metaproteins and coagulated proteins. Secondary proteins are those that are formed when much change has taken place due to further hydrolysis and some examples are proteoses, peptones, peptides *etc.*

3.6. Protein Synthesis

One of the astounding phenomena in physiology is synthesis of proteins. This is a highly complex *process that takes place by the command of the genes, i.e.*, hereditary materials.

The phenomenon of protein synthesis is in fact, much more mysterious in the *body of plants*. The higher forms of plants know how to take up nitrogen from the soil, which is the *basic ingredient* of a protein molecule. The lower forms of plants as well as some of the higher plants are even capable to pick up molecular nitrogen (N_2) from the air. It is observed that most of the higher plants take up nitrogen from the soil in *oxidized form*. But on taking up the oxidized form, *i.e.*, the *nitrate* (HNO_3) form, they reduce it step by step and produce the intermediate stages, which are, nitrite (HNO_2), hyponitrite (HNO), hydroxylamine (NH_2OH) and lastly, ammonia. This ammonia is utilized by them to make *amino acids* and the amino acids are used to synthesize protein molecules.

(It is observed that although the plants convert the nitrate form of nitrogen to the reduced form in their body because the reduced from in their body is necessary for synthesis of various amino acids, if both the oxidized and the reduced forms of nitrogen are present in the soil, majority of the plants prefer to take up the oxidized form of nitrogen, *i.e.*, nitrate).

Animals and the human beings are unfortunately *not capable* to take up nitrogen from the air or from the soil. Hence, *they solely depend on plants for nitrogen*. That is, they secure protein from plants or from other animals and on getting the proteins, they hydrolyze (= catabolyze) them to get the amino acids. However, they may also take up amino acids directly from plants or other animals. From the amino acids which the animals and human beings obtain either due to hydrolysis, *i.e.*, break down of protein molecules or directly from them, they are able to produce various types of amino acids. This is done by them by the process of interconversion, *i.e.*, *transposition* of the amino group of one amino acid into a keto acid. Following this process, the amino acids of various types that are required for protein synthesis are produced. For protein synthesis, 20 types or more of amino acids are required but by this process, all these 20 types are not produced by them directly in their body and for those amino acids which are not produced in their body, they depend on other source.

Protein synthesis takes place by the act of *joining*, *i.e.*, linking up of many amino acids. This joining (= condensation) process is carried out by the *direction, governance and participation of nucleic acids.*

In the present text, the mechanism of protein synthesis has been discussed only in brief. For detailed knowledge on this topic, the reader should consult other titles on physiology and biochemistry. However, keeping in view the fact that some knowledge on amino acids and nucleic acids are necessary to have an understanding on protein synthesis, these have been stated below before taking up the subject of protein synthesis.

3.6.1 Amino Acids

An amino acid is an ammoniacal organic acid which is composed of, (i) one *carboxylic group* (COOH) and (ii) one *amino group* (NH_2). Thus, it contains both acidic part (= carboxyl) and basic part (= amine) and hence, it is *difunctional* in nature.

The chemical formula of an amino acid is:

= Ⓡ - $CH(NH_2)$ COOH,

where [R] differs according to the type of amino acids. That is, R is the side-chain and it differs in accordance with types of amino acids. It is represented in Figure 8.

About 300 types of amino acids exist in nature. Among these, 20 types are required in protein synthesis. (However, this is controversial).

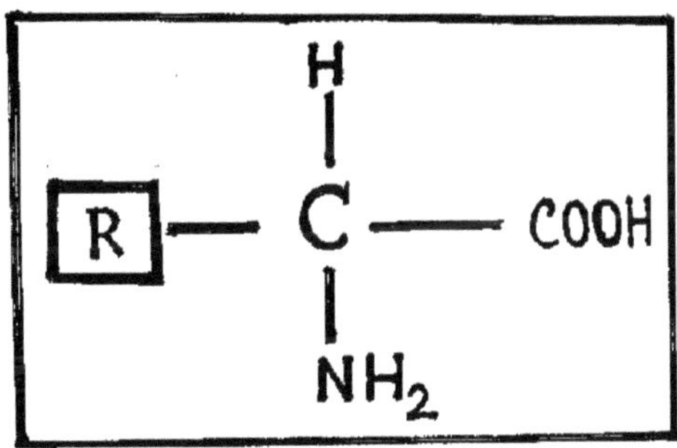

Figure 8. Structure of Amino Acid.
(R differs according to the type of amino acids).

3.6.1.1 Properties of Amino Acids

All amino acids exist in the form of alpha-amino acids. That is, the amine part of one acid is attached to the alpha-carbon part (substituent) which is next to the carboxyl group. They are soluble in water but insoluble in organic solvents. Mostly they are crystalline. As the amino acid molecules contain both acidic and the basic part, they form esters when reacted with either acids or alkalis. Due to acid-base reaction in the same molecule, the amino acid molecules exist primarily as *dipolar ions* (= hybrids that are known as *zwitterion*, derived from the German word, zwitter).

The amino acids have asymmetric carbon atoms and hence, they are *optically active*. That is, they exist either as dextrorotatory form or as laevorotatory form but the latter form is in greater existence. However, the simplest form of amino acid, which is *glycine* is not optically active. In an electric field, the amino acids migrate to the cathode with positive charge in acidic solution but this is opposite when they are in alkaline solution. Under certain conditions, they remain as neutral with no charge in electric field.

3.6.1.2 Classification of Amino Acids

The 20 types of amino acids that are involved to construct protein molecules are classified into three *broad groups*, *i.e.*, A, B and C and *sub-groups* belonging to these groups. These have been listed below:

(A) Aliphatic Amino Acids

(a) *Mono-amino, mono-carboxylic acids*

(i) Glycine (alpha-amino, acetic acid).

(ii) Alanine (alpha-amino, propionic acid).
(iii) Serine (alpha-amino, beta-hydroxy propionic acid).
(iv) Threonine (alpha-amino, beta-hydroxy, n-butyric acid).
(v) Valine (alpha-amino, isovaleric acid).
(vi) Leucine (alpha-amino, isocaproic acid).
(vii) Isoleucine (alpha-amino, beta-methyl-n-valeric acid).

(b) *Mono-amino, di-carboxylic acids :*
(i) Aspartic acid (alpha-amino, succinic acid).
(ii) Glutamic acid (alpha-amino, glutaric acid).

(c) *Di-amino, mono-carboxylic acids* :
(i) Arginine (alpha-amino, delta-guanadina-n-valeric acid).
(ii) Lysine (alpha-sigma-diamino-n-caproic acid).

(d) *Sulphur containing amino acids* :
(i) Cysteine (alpha-amino, eta-thio-propionic acid).
(ii) Cystine (alpha-amino, di cysteine).
(iii) Methionine (alpha-amino, gama-methyl, thio-n-butyric acid).

(B) Aromatic Amino Acids

(i) Phenylalanine (alpha-amino, beta-phenyl-propionic acid).
(ii) Tyrosine (alpha-amino, beta-para-hydroxy – phenyl-propionic acid).

(C) Heterocyclic Amino Acids

(i) Tryplophan (alpha-amino, beta-indole – propionic acid).
(ii) Histidine (alpha-amino, beta-imidazole – propionic acid).
(iii) Proline (alpha-pyrrolidine-carboxylic acid).
(iv) Hydroxy proline (4-hydroxy proline).

3.6.1.3 Synthesis of Amino Acids

Amino acids are synthesized by reductive amination in which organic acids are reduced by ammonia and more commonly by *transamination*. In transamination, the amine group (NH_2) in the alpha-position, except a few like lysine, threonine *etc.* is transferred to an alpha-ketoglutaric acid. By such transference, the ketoglutaric acid turns into glutamic acid and the amino acid from which the amine group is transferred is converted into a keto acid. This interconversion is carried out by the enzyme, amino acid transferase, *i.e., transaminase* (EC 2.6.1.21).

It may be stated that two transaminase enzymes are most active in the transamination process, These are, (i) glutamate–oxaloacetate transaminase, which acts on glutamic acid with oxaloacetic acid so as to produce alpha-ketoglutaric acid and aspartic acid, and (ii) glutamate–pyruvate transaminase, which acts on glutamic acid with pyruvic acid to produce alpha-ketoglutaric acid and alanine.

3.6.1.4 Essential and Non-essential Amino Acids

Plants of the higher and the lower groups are able to synthesize in their body all the 20 types of amino acids that are considered to be required for protein synthesis in them. But the animals and the human beings cannot synthesize by themselves all these 20 types of amino acids. It has been known that they are able to synthesize in their body only 12 types of amino acids but not the rest 8 ones. Hence, they are to depend on external source to get these 8 amino acids. From external source denotes that they are to take these 8 amino acids from the food which they take (= diet).

These 8 types of amino acids which they have to be got from outside, *i.e.*, diet. These are, threonine, tryptophan, isoleucine, leucine, lysine, phenylalanine, valine and methionine and these 8 amino acids are termed as *essential amino acids*. On the other hand, the 12 other types of amino acids which can be produced by themselves in their own body are regarded as *non-essential amino acids*. It may be mentioned that the amino acids, *viz.*, arginine and histidine were formerly considered as essential amino acids but now, these are not considered essential and a team of physiologists have preferred to group them as *semi-essential amino acids*.

3.6.1.5 Amino Acid Pool

The total amount of amino acids that are present in the body at any particular time is referred to as the *pool of amino acids*. The amount of amino acids that are required for protein synthesis are supplied by the pool when they are needed. The amino acids always come to the pool from various sources and also concomitantly go out of the pool for utilization. Nevertheless, size of the pool in the body does not change much by such coming or going of amino acids.

3.6.1.6 Amide

When an amino acid contains an *extra aminc* (NH_2) group, it is called an *amide*. That is, amide is an amino acid in which amine (NH_2) content is higher. Among other amides, the best known amide is *glutamine*, which is the amide of glutamic acid and is produced when *glutamic acid reacts* with *ammonia*. However, the amide, *asparagine* formed by the action of *aspartic acid* and ammonia is also of much importance.

Amide is chemically a rich source of amine group and thus, it can be said to make *storage of ammonia*. Accordingly when there is necessity of amino acids in the body, they break down chemically, *i.e.*, hydrolyze (= catalyze) by enzymes and liberate the enzymes for ultilization as amino acid synthesis.

Amides have however, *protective* role in the body also, which it does by trapping ammonia gas. Sometimes excess ammonia is produced in the body due to excess synthesis of amino acids within the body, or when it comes to the body from external source, *i.e.*, consumption of such food which is rich in protein. (It may be mentioned that some people eat too much amount of meat, egg, chicken and fish). Ammonia gas is highly toxic to the body and its excess accumulation is very harmful to some organs and particularly the brain. However, to get rid of the harmful effect of ammonia, the body converts ammonia into urea in the liver and then expels out the urea through urine. The process is known as *deamination*. Kidney also deaminates glutamine. Amide formation is actually a contrivance to *trap excess* ammonia, if acccumulated in the body. It may also be stated that although accumulation of *ammonia* is harmful to the body, it is not so if amide gets accumulated.

3.6.2. Nucleic Acids

Protein synthesis is carried out by the *direction* and *control of nucleic acids* and the operation is performed by joining, *i.e.*, linking (= condensation) of many amino acids. Nucleic acids are present in the *nuclei of cells* and hence, they are called nucleic acids. However, later it has been found that they are also present in some cytoplasmic organelles, *e.g.*, mitochondria, ribosomes *etc.*, but still, the name nucleic acid has not been changed.

There are two types of nucleic acids. These are called *deoxyribose nucleic acid* (abbreviated as DNA) and *ribose nucleic acid* (abbreviated as RNA).

3.6.2.1 Basic Chemical Configuration of Nucleic Acids

Basically, each of the two nucleic acids, *i.e.*, DNA and RNA is formed by some specialized chemical structures, which are known as *nucleotides*. Many nucleotides are joined together (= linked up) to form a nucleic acid. That is a nucleic acid (DNA or RNA) is actually a polymer of nucleotides, and the nucleotides are the monomers. Being polymers of nucleotides, a nucleic acid is also called a *polynucleotide*.

Each nucleotide is again, formed by joining (= linking up) three separate parts or units (= molecules). These are follows:

(i) A pentose sugar which is always a ribose.

(ii) A phosphoric acid.

(iii) A nitrogenous base.

Among the above three components, the ribose sugar is present in the centre. That is, it acts as the *central holder* and the two other components are linked up at the two ends of the ribose sugar. This signifies that the nitrogenous base is linked with the ribose sugar at one end, *i.e.*, at its outer side end which is at the No. 1 carbon atom of it. On the other hand, the phosphoric acid is linked with the ribose sugar at the opposite end, *i.e.*, at its inner end which is at the No. 5 carbon atom of it. The configuration is shown in Figure 9.

The nitrogenous bases are chemically, heterocyclic compounds. There are two types of these bases. These are called, *purine* and *pyrimidine*. Under the purine bases, *two* compounds are there, which are called *adenine* and *guanine*. On the other hand, under the pyrimidine base, *three* compounds are there and these are, *thymine*, *cytosine* and *uracil*.

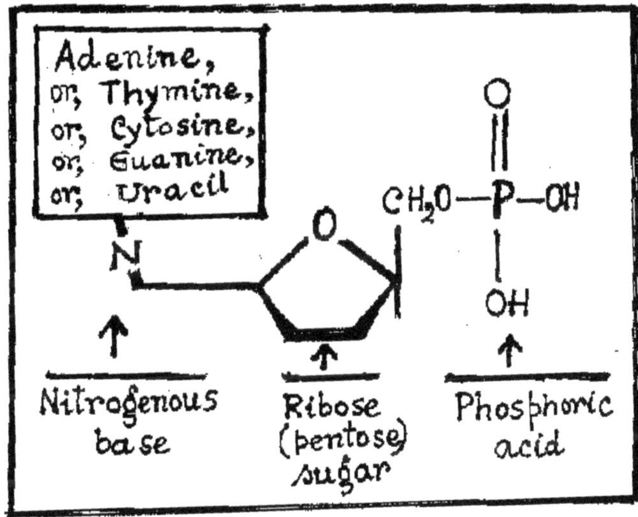

Figure 9. Chemical Structure of a Nucleotide.

The above configurations of nucleic acids are however, general. The actual configuration, *i.e.*, the chemical compounds of the nitrogenous bases as well as that of sugar differ according to the DNA and RNA. These have been stated under the next sub-heading.

3.6.2.2 Chemical Structure of DNA

Two pioneer scientists, *viz.*, J. D. Watson and F. H. C. Crick had jointly brought forward the chemical structure of DNA and hence, the model is called *Watson and Crick model*. The model suggests that DNA has actually a *double-helical, i.e.,* spiral like structure. This signifies that the DNA molecule has two thread-like strands, *i.e.*, filaments and these two strands remain in a *helical, i.e., twined manner around a common axis* (like a spiral star-case). Thus, one strand remains complementary to the other strand.

A nucleotide belonging to one strand is linked, *i.e.*, joined (= attached, bridged, bonded, interconnected) with the *complementary nucleotide* of the opposite (vis-a-vis) strand and this linking is done by means of *hydrogen bonds*. Hydrogen bonding takes place in such manner that the compound, adenine is linked, *i.e.*, joined *only with the compound, thymine* belonging to the pyrimidine base and that, this linkage is done by means of *only two hydrogen bonds.* In the same way, the compound, guanine belonging to the purine base is linked *only with the compound, cytosine* of the pyrimidine base but this linkage is done by *three hydrogen bonds* and not by two hydrogen bonds as in case of linking of adenine and thymine. This is shown in Figure 10.

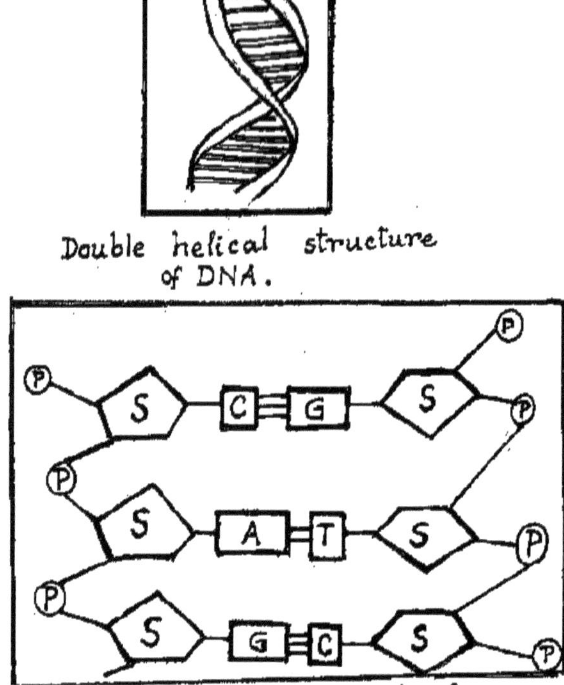

Figure 10. Arrangement of Nucleotides in DNA
(S: Sugar; C: Cytosine; G: Guanine; A: Adenine; T: Thymine; P: Phosphoric acid).

The above trend continues along the entire length of the spiral helix. Twisting of the two strands of DNA takes place in such a manner that there are about ten nucleotides in each strand for a complete turn.

As regards the sugar in the nucleotides of DNA, it is *always a de-oxy-ribose sugar*. Hence, it is called, *de-oxyribose nucleic acid.*

3.6.2.3 Chemical Structure of RNA

RNA differs from DNA in respect of the following three aspects in their nucleotides.

(i) Sugar in the nucleotide of RNA is always a *ribose and not a de-oxyribose* as in case of DNA. Hence, it is known as *ribonucleic acid* (= RNA).

(ii) The compound, thymine belonging to the nitrogenous base, pyrimidine is *not present in the RNA* strand. Instead of that, a different compound of pyrimidine base is present and this is known as *uracil.*

(iii) RNA has only one strand. That is, it is a single-helical structure and not double-helical like DNA strands.

3.6.2.3.1. *Types of RNA and their Characteristics*

Although DNA has only one type, RNA has tree types. These are termed as follows:

(i) *Messenger-RNA*, which is abbreviated as *m*-RNA. This is also termed as nuclear-RNA.

(ii) *Ribosomal-RNA*, which is abbreviated as *r-RNA*.

(iii) *Transfer-RNA*, which is abbreviated as *t-RNA*. Other terms of this RNA are, soluble-RNA (abbreviated as s-RNA), amino acid acceptor–RNA and adapter–RNA.

Apart from the above, there is another RNA, which is known as *viral-RNA* but this RNA is not concerned with protein synthesis.

Characteristics Features

(i) Messenger-RNA

This is formed, *i.e.*, originated from the DNA which is present in the nucleus (chromosome) of cell. In its formation, *i.e.*, synthesis of *m*-RNA, the two strands of the DNA are at first uncoiled, *i.e.*, they are *loosened apart (become free) from their double helical (spiral) form*. After this, one of the two strands acts as a *template*. (= By template is meant a mold, on which a structure is formed, *i.e.*, copied or duplicated). On this template, a *messenger*-RNA is formed, *i.e.*, copied. This process of copying is known as *transcription* and this transcribing reaction is catabolized by the enzyme, which is known as *RNA-polymerase* (Enzyme Commission No. 2.7.7.6). Since the *m*-RNA is formed by copying from a strand of DNA, obviously its base sequence remains *exactly same as with the strand of the DNA* from which it is copied.

Just after synthesis of the *m*-RNA has taken place, it leaves the nucleus and moves to the cytoplasm of the cell. At the cytoplasm, the *m*-RNA *comes to the ribosome* which is a cellular organelle and on which, the protein is synthesized, *i.e.*, this is the site of protein synthesis. Each *m*-RNA is *imprinted, i.e.*, marked (= stamped, engraved) with the nucleotides which carry the message (information) from the DNA. DNA is the *genetic material* which dictates (= gives instruction) as regards what protein(s) is to be synthesized. The *m*-RNA *transmits the message to the t-RNA* (= sends the message to the *t*–RNA) which is located in the ribosome (in the cytoplasm). Thus, the function of the *m*-RNA is to act like a *messenger to send the message from the DNA (= genetic material)* to the *t*-RNA.

(ii) Ribosomal-RNA

This RNA a (*r*-RNA) is present in the ribosome which is present in the cytoplasm of cell and which is the *site* (= location, platform) where protein is synthesized. Ribosomes usually exist in *clusters* and hence, when exist in clusters, they are known as *polyribosomes* (poly = many) or *polysomes* or *ergosomes* which have been stated later. In accordance with sedimentation velocity, a ribosome has two parts, *i.e.*, it has two subunits and there are known as 40 S and 60 S sub-units. It may be mentioned that exact role of *r*-RNA is not clearly known.

(iii) Transfer-RNA

This RNA (= *t*-RNA) is present in the cytoplasm of cell. Its function is to *selectively pick up the specific amino acid* which is actually required for protein synthesis and then *transferring that amino acid to the site of protein synthesis, i.e.*, polysome.

Transfer-RNA is a single-stranded nucleic acid like that of *m*-RNA or *r*-RNA and is not a double-helix like DNA. But interestingly, the *single strand folds back on itself at the upper end.* On doing so, it falsely gives a double-helical, *i.e.*, double-stranded appearance like DNA, although in real sense, it does not contain two separate strands. However, as it makes two strands by folding, one strand of it acts as *complementary to the other strand* (vis-a-vis). Accordingly, the *nucleotides of one strand links up, i.e., joins with the corresponding nucleotides* that are present in the complementary (vis-a-vis) strand. This looks like pairing of the two strands of DNA.

The joining of nucleotides takes place in such a manner that the compound, *guanine* belonging to the purine base is linked (joined) with the *cytosine* of the complementary, *i.e.*, oppsite strand (which is similar to DNA). However, the compound, *adenine* of the purine base is linked with another compound, which is known as *uracil* beloning to the pyrimidine base. In fact, in the strand of *t*-RNA, thymine of the pyrimidine base is not present and *in its place, uracil is present.*

It should be noted that when the same strand of the *t*-RNA folds back on itself, at the turning end of it, *i.e.*, at the head-end, three nucleotides are present side by side. These three nucleotides remain as free, *i.e., they are not ioined with any other nucleotide.* These three free, *i.e.*, unpaired nucleotides are *collectively called anticodon,* or *anticodon triplet.* This shows that the anticodon triplet is positioned just opposite to the triplet which is recognized as codon, that is located in the *m*-RNA. Thus, on such positioning, the set of *codon in the m-RNA is placed just opposite to (vis-a-vis) the set of anticodon triplet in the t*-RNA.

The *m*-RNA has series of nucleotides. These are arranged side by side and remain unpaired, *i.e.*, not joined (linked) with any other nucleotide. These nucleotides of the *m*-RNA have the function to take up (= carry) information from the DNA, *i.e.*, the genetic material as to what type of proteins have to be synthesized. Thus, the *nucleotides of the m*-RNA *encode secret information obtained from the DNA as regard what proteins have to be synthesized, i.e.*, DNA dictates the nucleotides of the *m*-RNA in regard to what proteins that are to be synthesized. (When three of the nucleotides lying side by side, whose base sequences are exactly complementary to those of the nucleotides of the anticodon of the *t*-RNA, these three nucleotides together is called a *codon*).

In order to synthesize protein molecule, the *nucleotides of the t-RNA selectively pick up the necessary amino acids* which freely float in the cytoplasm. On picking up the required amino acid molecule, the nucleotide moves on along with the amino acid and comes close to the position of the *correct (nucleotide of the) m*-RNA. There it takes position in such a manner that the anticodon triplet of the *t*-RNA exactly fits, *i.e.*, matches up to the codon of the *m*-RNA. That is, *the anticodon triplet of the t*-RNA *is positioned to the complementary codon of the m*-RNA. (This may be understood on consideration that in a torch light, only when the positive pole of a battery touches the negative pole of the adjacent battery, the bulb of the torch light glows because of having electrical connection).

At the opposite end of the 'head' of the *t*-RNA, *i.e.*, at its tail-end, there are two opposite ends which remain open, *i.e.*, free. At one end of these two free-ends, *three nucleotides* are present and their base requences are C-C-A. At the other open end of the tail, only *one nucleotide* is present and this nucleotide is G. The three nucleotides at one open-end, which are C-C-A as well as the one nucleotide at the other open-end, which is G are *not paired with any other nucleotide, i.e.*, they remain free. During protein synthesis, the selected amino acid *gets attached to the t*-RNA at the C-C-A end only. The configuration of *t*-RNA as presented by Holley is given below in Figure 11.

3.6.3. Mechanism of Protein Synthesis

A protein molecule is very large in structure and highly complex. It is composed of plentiful amino acids which may be more than a hundred in number.

The entire process of protein synthesis is very complicated. This involves a number of *steps* or *stages* and these have been briefly outlined in the following:

Step 1 : Activation of the Amino Acids

The amino acids with which a protein molecule is to be formed floats freely in the cytoplasm. These amino acids are actually in *inactive form*. Hence, these are *needed to be activated at first and only when they are activated, they could be utilized for protein synthesis.*

Activation is the process by which a material is made potentially active to undergo chemical reaction and for this, another material is needed to be added to it. An amino acid becomes active *when it is added, i.e., reacted with ATP*. An enzyme, known as *amino acyl synthetase* catalyzes this reaction, along with a metallic co-factor, which is magnesium. As a result of this enzymic reaction an *amino acid complex* is formed and this is known as *amino acid adenylate* (= amino acyl adenylate). Besides this adenylate compound which is formed, the compound known as *pyrophosphate* is released in this reaction but this is not used.

When the amino acid turns to this adenylate complex, it becomes activated. The activated amino acid is then capable for protein synthesis. Hence, it is picked up by a specific *t*-RNA for protein synthesis. When the *t*-RNA picks up the activated

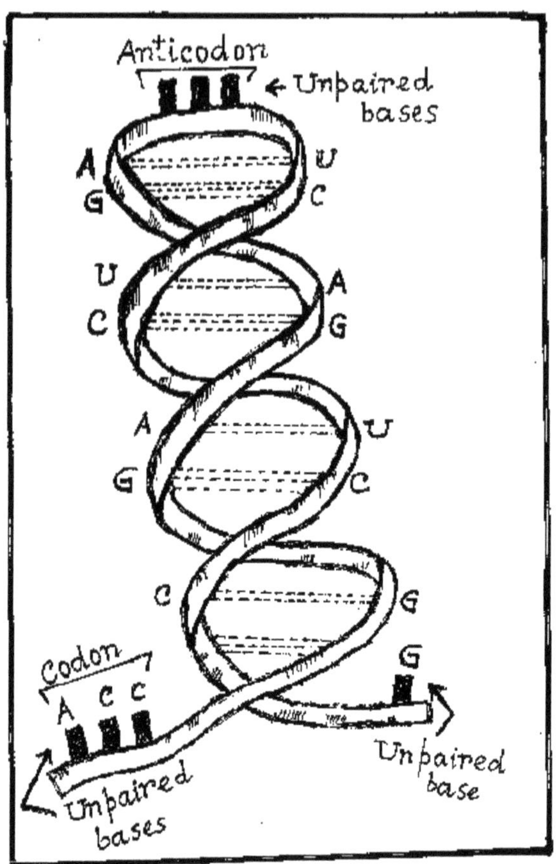

Figure 11. Structure of Transfer-RNA
(Single-helical-strand folds back to assume double-helical appearance).

amino acid, it is said to be a *charged t-RNA*, or, amino acid loaded *t*-RNA. This charged *t*-RNA then moves to the site of protein synthesis and gets attached with the *m*-RNA. But the enzyme, amino acyl synthetase till then remains attached with the amino acyl adenylate. This is because, during transfer of the amino acid to the *t*-RNA, this enzyme catalyzes this process such that the adenylate part is detached and only the amino acid remains attached to it.

An important consideration in the above reaction is *specificity*. That is, firstly a *t*-RNA *is specific only for a specific amino acid.* This means that, a specific *t*-RNA can only pick up a specific amino acid (activated) and not any other amino acid. Secondly, *for each amino acid, a specific synthetase enzyme is required.* Therefore, 20 types of amino acids that are required in protein synthesis, there must be at least 20 amino acyl synthetase enzymes.

Step 2 : Attachment of *t*–RNA to the Ribosome

Protein synthesis takes place on the ribosome and the process is directed by the *m*-RNA. Hence, the *m*-RNA, (which is actually *encoded with the genetic information received from the DNA* in regard to what protein is to be synthesized) comes out of the nucleus and reaches the cytoplasm of the same cell. In the cytoplasm, the *m*-RNA reaches out the specific ribosome where protein is to be synthesized and it gets attached, *i.e.*, fixed there. A ribosome has two parts, of which one is smaller in size than the other. These are called *sub-units*. One sub-unit is 40 S and the other is 60 S in size. The *m*-RNA gets attached to the ribosome at its 40 sub-unit only.

Step 3 : Transference of the Charged Amino Acids

When all the amino acids which are to be used in protein synthesis have been eharged, i.e., activated, a particular amino acid which is activated is picked up by a specific *t*-RNA. The *t*-RNA also contains the anticodon triplet which is present at its head-end (Figure 11) finds out the *particular part of the m-RNA, i.e.*, the codon triplet of the *m*-RNA and there, it properly fits. That is, the three nucleotide bases of it that acts as *anticodon triplet* corresponds (= complementary to) with the group of three complementary bases of the *m*-RNA acting as codon. Thereby, the anticodon triplet of *t*-RNA properly fits to the codon triplet of the *m*-RNA (vis-a-vis). For example, if the base sequence of the anti-codon in the *t*-RNA is say, UAC (Uracil-Adenine –Cytosine), then the base sequence of the codon in the *m*-RNA should be AUG (= Adenine-Uracil-Guanine).

On coming into such position, the compound, amino acid adenylate is *transferred to the t*-RNA. The enzyme, amino acyl synthetase (Enzyme Commission No. 6.1.1) which remains present catalyzes the transfer reaction. After the amino acid is transferred, the enzyme and the ATP are detached. Therefore, *only the amino acid and not the adenylate is transferred*.

An important aspect should be noted in this connection. That is, an *m*-RNA is formed on transcription, *i.e.*, duplication from one of the two strands of the DNA. Therefore, the nucleotides that are present in the *m*-RNA must have to be in conformity with the nucleotides from which they have been copied, *i.e.*, the strand of the DNA. The DNA is genetic material and its function is to *dictate message* and this is done by symbols in other words, *codes*. The codes get imprinted (= impregnated) in the *m*-RNA when it is synthesized, *i.e.*, formed (=copied) from the DNA. These codes which are inscribed in the *m*-RNA through the nucleotides of it are called *genetic codes*. This is because, the nucleotides *encode the informations* from the DNA as regard what proteins are to be synthesized.

The nucleotides which are present in the *m*-RNA in turn, transfer (= send forth) the information to the corresponding *t*-RNA. On receiving the information, *i.e.*, message from the *m*-RNA, the *t*-RNA picks up the specific amino acids from the cytoplasm that have been already activated. Thereby, the specific protein molecules are synthesized.

It is thus, clear that specific proteins are synthesized as per direction, in other words, *dictation of the genetic materials, i.e.,* DNA molecules, located in the chromosomes in the nuclei of cells and this is carried out by the DNA molecules through specific codes.

Step 4: Detachment of *t*-RNA

When the activated amino acids are transferred to the ribosomes, they synthesize protein molecules and the *t*-RNAs which have carried them there become free, *i.e.,* detached. The free *t*-RNAs then come out and search other activated amino acids in the cytoplasm for the purpose of doing other proteins. In this way, proteins of various types are formed by the act of joining amino acids which is brought about by the *t*-RNAs.

In the mechanism of protein synthesis, it should be remembered that the ribosome, which is the site of protein synthesis, *i.e.,* where the *m*-RNA gets attached in the cytoplasm does not occur singly. In fact, the ribosomes exist in *groups, i.e., in clusters* of 3 to 6 or more. *Such a group of ribosomes is called a polysome, i.e., polyribosome* (poly = many) or is also called ergosome. Protein synthesis takes place on a polysome and not on a single ribosome. The group (cluster) of ribosomes (= polysome) are actually *interconnected* and this is done by the strand of the *m*-RNA. Accordingly, a number of ribosome groups remain attached to the same m-RNA strand, *i.e.,* they are *arranged in a line.*

When a protein molecule is synthesized, the group of ribosomes (= polysome) moves (= slides) away to the side along the line of the *m*-RNA. In this way, when the ribosome group comes to the place, *i.e.,* at the place (location) which has been made free by the first group of ribosomes. Lastly, the ribosome groups become *separated, i.e.,* they are detached from the *m*-RNA strand. Then the protein molecules formed are also freed, *i.e.,* detached. On detachment, the detached proteins are released and they remain in the cytoplasm. The detachment of a protein molecule takes place *very shortly* after it is synthesized and the process is continued to synthesize other proteins. The formation of a protein molecule by joining of amino acid molecules has been represented in Figure 12.

It is now confirmed that the DNAs are present in the chromosomes located in the nuclei of cells and these are the genetic materials. But later it has been known that DNAs are present not only in the chromosoms but are also present in the *mitochondria, chloroplasts, nucleoli* and may be in other cellular organelles located in the cytoplasm. The mitochondria (singular : mitochondrion) are present in the cytoplasm of both animal and plant cells and this organelle is the seat of respiration in the body. The organelle, *chloroplast is present only in green parts of plants* and is the seat of carbohydrate (glucose) synthesis. It seems apparent that as these organelles contain DNA, they are also capable of protein synthesis. Experiments have supported this fact but the amount of protein that is produced by these organelles are very low as compared to the amount of protein, synthesized by the organelle, ribosome (= polysome).

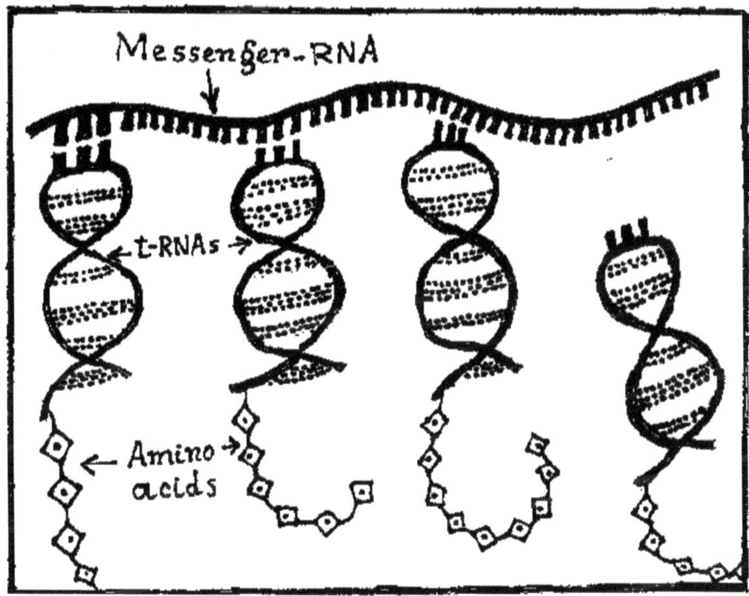

Figure 12. Synthesis of Protein Molecules by Joining of Amino Acid Molecules.

3.7. Breaking Down of Protein Molecules

Breaking down, *i.e.*, catabolism of a protein molecule is its enzymatic hydrolysis. The pathway of break down appears to be the reverse of the synthetic pathway. The intermediate products that are successively produced during break down of a protein molecule are *proteoses, peptones, polypeptides, tripeptides, dipeptides* and *amino acids*. From theoritical viewpoint, when any particular protein molecule is catabolyzed (= hydrolyzed, broken down), the same types of amino acids and same amount of each of them with which the particular protein molecule has been formed should be obtained. But from the practical point of view, this does not happen. In fact, there would be competition between the intermediate products that are formed during hydrolysis. Besides that, some of the intermediate products are quickly ultilized in the synthesis of new protein molecules. Studies on degradation of protein molecules have according to Webster not been so adequately studied as compared to the synthetic mechanism of protein molecules.

Mention may be made in this connection that from the nutritional point of view, broken down products, *i.e.*, hydrolysates of proteins are much more effective than the proteins which are not hydrolyzed. This is because, the hydrolysates (= intermediate products) *ensure quicker availability* than the non-hydrolyzed protein molecules which are of highly complex nature. In this regard, the shorter chain intermediate products are much more effective and amino acids are most effective and most quick acting.

Protein hydrolyzing enzymes are proteases, derived from the *pancreas gland* (= pancreative protease) and *trypsin, chymotrypsin etc.* derived from the intestine.

Proteolytic activity is observed in some plants, *e.g., papain, chymopapain, papaya oxidase etc.* These are obtained from the white milky latex of young papaya fruits and also leaf veins, *bromelain* from the fruit and stem of pineapple, *ficin* from the fruit of fig, *actinidin* is obtained from the kiwi fruit *etc.* Protein hydrolyzing activity is also observed in some bacterial species *e.g., Bacillus sublitis, B. licheniformis, Pseudomonas arguinosa etc.* and also from some fungal species.

In regard to hydrolysis of protein inside the body, some endogenous fators are involved. An example is *thyroid hormones* which are essential in synthesis of proteins but high level of the same may cause break down of protein molecules.

3.8. Regulation of Protein Molecules

Protein metabolism is influenced by a host of factors. *Thyroxin* (= 3,5–3'5' –tetra–iodo–thyronine), which is one of the principal hormones secreted by the thyroid gland significantly increases the rate of protein synthesis in the body. Protein metabolism is also affected by *adrenocortical hormones,* secreted by adrenal cortex. Above all, the type, quality and amount of protein taken through food has a marked influence on the protein content of the body.

3.9. Digestion

A number of proteins, *e.g.,* albumin, globulin, caseinogen, collagen, gelatin, mucin *etc.* come to the body through food. These are highly complex in chemical structure and of high molecular weight, for which their absorption in the body becomes difficult. Hence, the body *digests* them, *i.e.,* chemically breaks down (= catabolyze), into *short-chain peptides* and ultimately, into amino acids.

Digestion of proteins actually starts in the stomach by reaction with *pepsin,* which is formed from pepsinogen by the action of hydrochloric acid. Pepsin acts in acidic medium (at pH 1 to 3) and as the stomach contains sufficient hydrochloric acid, the process is facilitated by this acid. Due to digestion by pepsin in the stomach, the protein molecules are broken down (= catabolyzed to the intermediate compounds, which are, metaprotein, first and second proteose and peptone. (Pepsin is endopeptidase and hence, acts at such points which are in the interior part of the polypeptide chain). Digestion of protein molecules liberates amino acids but to a small extent. An important protein, collagen is however, digested by the action of acidic-pepsin enzyme.

From the stomach, the undigested or partially digested preteins are passed on to the small intestine. Inside the lumen of the small intestine, the protein molecules are reached by the enzymes, *pancreatic protease* (EC No. 3.4.21.70), trypsin, chymotrypsin, elastase, carboxy-peptidase *etc.* These enzymes catalyse those proteins which have not been hydrolyzed by pepsin, or only partially hydrolyzed. Thus, catalytic function of pancreatic protease has greater significance than the pepsin. However, on being hydrolyzed by pepsin and more so, by the pancreatic enzymes, protein molecules are gradually broken down (= digested, catabolyzed) sequentially into proteoses, peptones, tripeptides, dipeptides and ultimately, amino acids.

3.10. Absorption

Absorption of amino acids, dipeptides, or tripeptides takes place in the body mainly by *secondary active transport* system and it is carried out mainly in the jejunum, although may be to a small extent in the ileum. This is brought about by a protein known as *carrier protein*. The carrier proteins are of various types and each one of them has two opposite sites. At one site, it couples (= binds) with sodium (Na^+) ions and at the other site couples with the respective amino acid, or dipeptide, or tripeptide as the case may be, *i.e.*, which is needed to be transported. The *carrier-proteins* (= carrier-mediated proteins) are different for the different peptides and their amino acids. In other words, they are *specific, i.e.*, separate for dipeptides, tripeptides and amino acids, which are of basic or neutral reaction. In this connection, it may be mentioned that the amino acids that are of *laevo* form are usually more readily absorbed than those which are of *dextro* from and the latter forms are absorbed through passive absorption.

Some proteins are directly absorbed in the body, *e.g.*, in the case of suckling infants. The protein of the mothers' milk is directly absorbed in their intestines and due to that, the *infants get the antibodies directly from their mothers' milk*. Some people show allergic symptoms soon after taking those food to which they are hypersensitive. The antigens from those food are directly absorbed and reactions take place quickly.

3.11. Quality of Proteins

All proteins are not of same quality in respect of their amino acid profile and other attributes. Accordingly, nutritive value and other health-giving properties of them also differ.

What is actually meant by the quality of protein that is obtained from a food is in fact, difficult to be stated precisely. However, the nutritional scientists have considered a number of criteria to judge their quality and efficiency. Nevertheless, each of the assessment could be said to have some discrepancy. The type of amino acids and the quality of each that are present in a given protein is evidently, paid emphasis to judge its quality and in consideration to that, a protein sample which contains all the *essential amino acids* is designated as a " first class protein. "On contrary, a protein which lacks even one of the essential amino acids is graded as a second class protein." Almost all proteins that are derived from animal source food are considered as first class proteins and among exceptions, *gelatin* may be cited.

(a) Classification in Consideration to Growth in the Body on Consumption

This system primarily considers rate of growth in the body that takes place on consumption of a protein of particular amount that is derived from a given food. To judge the quality of proteins, three grades of them are made as follows:

(i) Complete Protein

Those proteins which make *best growth* in the body are ranked as complete protein. All the essential amino acids that are required in the body for ideal growth are in fact, present in the proteins belonging to this group.

(ii) Partially Complete Protein

Proteins of this group make *moderate growth* in the body. In fact, some of the important amino acids, *e.g.*, lysine and threonine are not present in these proteins for which they do not make similar type of growth as would happen on consumption of complete protein.

(iii) Incomplete Protein

Proteins under this group *do not promote almost any growth* in the body on consumption and if they do so, only scantily. These proteins do not contain a number of important amino acids. The gelatin protein (= zein) may be cited an example of this group.

(b) Classification in Consideration to some Specific Standards

Owing to some discrepancies met with in the above systems of grouping, the nutritional scientists have made groupings of different types in consideration to a number of attributes of the proteins. However, a number of classifications of proteins have been done by different authorities in consideration to specific standards and some are stated below, which are used now to assess the quality of proteins.

(i) Biological Value (BV)

Among other methods, determining quality of proteins in the food samples in consideration to biological value, *i.e.*, efficiency of them has received much attention. The amount of protein in the food that is actually utilized to meet the *requirement of protein in the body* is considered in this method. Rats used as experimental samples are either fed with protein-free diet or with ten per cent protein for a period of ten days. The excretory products of them, *i.e.*, urine and faeces are collected and the nitrogen content of those are determined. From that, the biological value of the protein sample is worked out.

(ii) Digestibility

How easily and quickly the protein is digested in the body is an important criterion to determine its quality. *Protein derived from milk and egg are known to be much more easily digested* and the time required for their complete digestion till breaking down to amino acids, *i.e.*, the last step of digestion is only 1–2 hours. On the other hand, proteins derived from pulses and some other vegetarian food take long time to be digested.

The reason for this delay is, many vegetarian food contain such substances that *inhibit the action of trypsin*, which is a notable protein-hydrolyzing enzyme in the intestine. However, cooking, boiling and roasting destroy, *i.e.*, inactivate the trypsin-inhibiting substances to a considerable degree. But protein of soybean (soybean contains 43.2 per cent protein, which is highest of all the plant and animal proteins on Earth) is not digested by cooking, boiling or heating. Heating food to a high degree has again, the demerit in that, it destroys some amino acids in the food, particularly lysine and methionine.

(iii) Protein Efficiency Ratio (PER)

The body weight increase by intake of protein is taken into consideration in this method. Like biological value, experimental rat samples are fed with protein-free food or food containing ten per cent protein for a certain period, after which the body weight of them is determined. The increase of body weight by feeding protein diet is considered and the efficiency of protein is determined. For some drawbacks, the PER is however, replaced by Relative Protein Value (RPV) of a food sample.

(iv) Net Protein Utilization

Nitrogen content in the body that is retained on taking protein from a given food is considered in this method. A direct method of determining Net Protein Utilization has been developed by Miller and Bender (1955) in which, 28-day old experimental albino rats were used. A group was fed with non-protein diet and the other group was given ten per cent protein from different types of food. After giving such diets for ten days, the rats of both groups were killed. After that, their dead bodies were dried and ground. The nitrogen content of the ground material of both groups was estimated and in consideration to these data, NPU was determined.

(v) Chemical Score

In this method, protein derived from egg is considered best and is given *a score mark of 100*. Protein of other food-stuffs is compared with egg-protein and is given score marks accordingly.

Swaminathan, P. in his book, "Essentials of Food and Nutrition – Volume I," 1974 (Ganesh and Co., Madras – 17) has stated the quality of protein of some food-stuffs, which is given in the Table 5.

3.12. Nutritional and Other Health Benefits

Functional processes in the body cannot be thought of without proteins. Next to water, protein is most abundant compound in the body. It is the structural and functional element in every cell and is the component in imparting structure of the cell membrane. Proteins function as regulators of *fluid balance* and *acid-base balance* and help the body to prevent from swelling, drying up and becoming toxic. It is a component of *enzymes* and *hormones* which regulate a great many metabolic processes. Antibodies and immunoglobulins which neutralize or destroy foreign (= exogenous) or auto-antigens in the body and thereby, protect the body against many infections. Proteins are responsible for synthesis of prothrombin, fibrinogen and other plasma proteins.

Table 5. Quality of Protein of some Food-stuffs
(Represented from Swaminathan, P., 1974 : Essentials of Food and Nutrition)

Food	*PER*	*BV*	*NPU*	*Chemical Score*	*Limiting Amino Acids*
Animal foods					
Egg	4.5	96	91	100	Nil
Milk, cow's	3.0	84	75	65	Sulphur Amino acids
Liver	2.9	77	65	66	Sulphur Amino acids
Meat	2.8	80	76	70	Sulphur Amino acids
Fish	3.0	85	72	60	Tryptophan
Cereals					
Rice	2.0	64	57	60	Lysine, threonine
Wheat	1.7	58	47	42	Lysine, threonine
Pulses					
Bengal gram	1.7	58	47	44	Sulphur amino acids
Peas, dried	1.6	56	45	42	Sulphur Amino acids
Nuts and Oilseeds					
Groundnut	1.7	54	45	44	Lysine, sulphur amino acids and threonine
Soybean	2.0	64	54	57	Sulphur amino acids
Cottonseed	2.1	63	53	55	Sulphur Amino acids and lysine
Coconut	2.5	67	56	52	Sulphur Amino acids, lysine and threonine
Sesame	1.7	60	51	40	Lysine

PER: Protein Efficiency Ratio; BV: Biological Value; NPU : Net Protein Utilization.

For synthesis of the globin part of *haemoglobin* in the blood, proteins and some amino acids, notably histidine, leucine, phenylalanine *etc.* have essential role. They act as *transporters* of many substances across cell-membrane. The bile acids, taurocholic acid and glycocholic acid are formed from amino acids.

Necessity of protein is felt in synthesis of *glutathione*, which is the tripeptide of glutamic acid, cysteine and glycine and *cytochrome*, the *haemochromogen* (= a compound of heme) which are necessary for exudation of cells.

Proteins have essential role for *growth and repair of tissues and muscles* in the body. The body tissues that break down during metabolism are repaired by proteins. In order to accumulate muscular mass in the body, proteins are indispensable and they constitute about 43 per cent of the muscles. Muscular contraction is regulated by the contractile proteins, *myosin* and *actin*. For storage of many substances, notably *ferritin*, the iron-phosphorous-protein complex, protein have their necessity.

Table 6. Protein Quality of some Food-stuffs

Protein Type	*PER*	*NPU*	*BV*	*Protein Digestibility (PD)*	*Protein Absorption Rate*	*Amino Acid Score (AAS)*	*PDCAAS*	*DIAAS*	*Limiting Amino Acid*
Whey	3.2	92 per cent	104	N/A	8-10 g/h	N/A	1.0	0.973-1.09	None
Casein	2.5	85-92 per cent	77	95.1-97.6 per cent	6.1 g/h	1.19	1.0		None
Egg	3.9	94 per cent	100	97 per cent	1.3-2.8 g/h	N/A	1.0		None
Beef	2.9	73 per cent	80	N/A	N/A	N/A	0.92	N/A	Methionine + Cystein
Soy	2.2	61 per cent	74	86 per cent	3.9 g/h	N/A	0.91-1.0	0.898-0.906	Methionine + Cystein
Rice	N/A	N/A	N/A	N/A	N/A		0.50		Lysine
Pea	N/A	N/A	N/A		2.4-3.4 g/h		0.597-0.70	0.579	Methionine + Cystein
Hemp	N/A	N/A	87	94.9 per cent	N/A	0.64	0.61	0.579	Lysine
Mycoprotein	N/A	N/A	N/A	86 per cent	N/A	N/A	0.996		Methionine + Cystein
Spiruline	1.8-2.6	53-92 per cent	68	83-90 per cent	N/A		N/A		None
Chlorella	N/A	N/A	N/A	N/A	N/A		N/A		Methionine + Cystein

A distinguished feature of protein is to serve as an *alternative source of energy* which it does on decomposition (= breaking down) of amino acid molecules. It is observed that 1 gram of protein gives rise to 4.2 Kcals (= 17.57 kilojoules) of energy. From calorimetric studies, it is known that energy value of protein is 5.3 Kcals per gram of protein. (A report of the World Health Organization states that protein contribute to 10–15 per cent of daily calorie intake).

In this connection, it should however, be brought to notice that *this is not desirable under normal condition of the body*. This is because, protein is a highly valuable substance as it has many important functions in the body. Using this compound to meet the requirement of energy is a sheer *misuse*. For the purpose of energy requirement, carbohydrate is the ideal material and 1 gram of carbohydrate gives the same amount of energy as 1 gram protein gives. Fat is still a more efficient source as 1 gram of it gives rise to 9 Kcals (= 37.8 kilojoules) of energy. Protein is used in the body to meet the requirement of energy when there is deficiency of carbohydrate or fat. Breaking down of protein molecules to liberate energy has another harmful effect in that, due to breakdown of protein molecules, ammonia gas in liberated and this gas has toxic effect in the body.

Accordingly, in order to meet up the requirement of energy in the body, protein should not be used, *i.e., protein should not be misused*. Instead of that, carbohydrate and if necessary, fat should be used. Carbohydrate and fat should in fact, *spare* the protein from its use to liberate energy. Hence, these are nicknamed as *protein sparers*. It should be remembered that there should be adequate carbohydrate and fat in the diet as in such case, the amount of protein that will be required to maintain *nitrogen balance* could also be required.

Nutritional deficiency of dietary protein results in a number of disease conditions. Among these, *kwashiorkar* and its serious form, *marasmus* in children are of particular mentioning. These are known as protein and/or, *protein-energy malnutrition syndromes* which were formerly termed as protein-calorie malnutrition of children. The word, Kwashiorkar has been adopted to the protein-starved disease by the people of Gold Coast country in Africa and it was introduced by Cicely Williams in 1935. Kwashiorkar is a severe deficiency disease of protein of the children in some parts of western Africa. Literal meaning of the term is, suffering of a baby when the mother gives birth to another body. In fact, the baby suffers from protein adequacy, *i.e.*, mother's milk when the next baby is born who is deprived of mother's milk. The abnormality is particularly prevalent in the poverty-stricken families as they cannot afford to purchase proteinaceous food for the baby. Syndromes rather than symptoms of this protein deficiency disease are retardation of growth, oedema in the legs, thighs, face and other parts of the body, wasting of muscles and a conspicuous symptom is, well-rounded shape of the face which is known as *moon-face*. Enlargement of liver, anaemia, dermatosis, angular stomatis, low level of serum albumin, alkaline phosphatase, lipase, amylase, fluid imbalance *etc.* are other symptoms. Besides these, mental behaviour of the baby is also impaired.

Kwashiorkar appears to a baby due to *severe deficiency of protein* and the related disorder, *marasmus* appears to a baby when it is *extremely deprived of both energy as well as protein*. Nutritional marasmus very commonly appears in the weaned infants of 1-year old unlike Kwashiorkar while Kwashiorkar appears when the infants attain 2–4 years of age. Symptoms of marasmus are extreme retardation of growth, wasting of muscles, dryness of skin, anaemia, deficiency of vitamin A, abdominal distension, diarrhoea *etc.* but oedema is hardly noticeable in marasmic condition. The intermediate form between Kwashiorkar and marasmus is called, "Marasmic-Kwashiorkar". The two abnormalities interchange according to circumstances. For example, an infant in early stage of Kwashiorkar may develop marasmus if suffers from infective diarrhoea and under-feeding. Again, an infant with marasmus may develop Kwashiorkar when it is fed on such food which is deficient in protein but rich in carbohydrate and adequate salt.

Importance of the protein, *collagen* in maintaining health needs no emphasis. This strong, fibrous, water-insoluble protein helps in providing *structural scaffolding* to the cells and is vital for *skeletal health*. It acts like a cushion so that end of a bone does not rub with the end of the adjacent one. Collagen has also a great role in maintaining integrity of tendons, ligaments, peridontal ligaments and in preventing against osteoporosis. Inadequacy of collagen is a cause of wrinkled skin and sagged throat and cheek. The protein, *insulin* which is the hormone secreted by the beta-cells of Langerhans of the pancreas has its indispensability to stabilize blood sugar and thereby, to ward off hyperglycaemia.

Protein is a building material for skin which contains about 15 per cent of the skin. The protein, collagen and elastin have vital roles in maintaining skin health and youthful skin. Collagen provides strength to the skin and elastin serves as the main component of the elastic fibres. Hair and nails in the body are largely made up of the highly tough protein, *keratin*, which is produced by keratinocytes. While the hard keratin is present in hair and nails, soft keratin is located in the skin epidermis.

Rhodopsin (= scotopsin) which is the visual purple pigment of the eye is formed by the combination of the protein, opsin with retinal, the vitamin A aldehyde of retinene. It is claimed that proteins play vital role for smooth functioning of the nervous system and formation and maintenance of myelin sheath. Protein-rich diet is also believed to promote restful sleep, possibly by achieving neurotransmitter (norepinephrine, acetylcholine, dopamine *etc.*) balance.

A section of nutritionists is of the opinion that protein has a role to combat obesity. How the protein acts is not clearly known but suppressing appetite and thermogenesis, *i.e.*, the necessity to expand more energy to digest food had been stated to be possible causes. It is known from a study done at the Harvard School of Public Health in U.S.A. that the subjects who took diet, rich in protein could be able to make better management of weight than those who consumed low protein diet. Benefit of high protein diet on cardiovascular health has also been indicated from a study undertaken at the John Hopkins University Medical School in the U.S.A. There is report that protein has a role even to promote longevity.

Formerly, protein was highly valued for its definite role in growth and wear and tear of the body, building up of muscles, repair of body cells, wound healing and also to provide energy under certain conditions. However, in course of time, its role for other benefits in the body and beneficial effect on certain illnesses has also come to knowledge. Hence, this invaluable organic compound is now regarded as a worthy *nutraceutical* (= nutritive + pharmaceutical) material rather than a nutritive component only. Nutritionists and other health scientists are of late, paying much attention to undertake further research studies to aim at exploring what are the diversified benefits that proteins give us.

3.13. Requirement in the Body and Recommendation

How much and what type of proteins the human body needs is difficult to be assessed and prescribed with exactness as a number of factors are involved on the path. Some of the pertinent factors in regard to proteins may be stated as lack of adequate knowledge on these compounds, their diversity, stabilization, denaturation, mode of action and particularly, lack of adequate experimentation on human subjects. In regard to making any recommendation, age of the individual, gender, body weight, general health, pregnancy, lactation, physical activity, illness if any, occupation, living condition of the subjects and the environmental factors need heedful consideration. Hence, a straight bat recommendation is difficult to give. The recommendations that have been provided by the nutrition scientists should be considered as general, *i.e.*, broad and to meet the actual requirement of the individual, these are needed to be somewhat modified.

Requirement of protein in the body has been determined by the scientists, sometimes in consideration to the requirement of specific amino acids in the body but in most cases, the requirement of protein in the body, its quantity and quality and the food-stuffs from which procured have been considered.

M. Swaminathan (1974) in his book, "Essentials of Food and Nutrition–Volume–I (Ganesh and Co., Madras - 17)" has presented the table, stating the consolidated reports of the *requirement of amino acids of various subjects* which is given in Table 7.

To meet the need of amino acids, the quantitative value for *egg protein* and *cow's milk protein* (gram per kg per day) have been stated as follows: Infants : 1.6 and 2.0 gram. Children (10 - 12 years) : 0.9 and 0.9 gram. Adults : Women - 0.18 and 0.28 gram. Adults : men - 0.26 and 0.43 gram.

The requirement of protein for an individual is determined by a number of methods. Among these, the more commonly followed methods are, (i) nitrogen balance, (ii) obligatory loss of nitrogen and the (iii) factorial method.

Table 7. Minimum Daily Requirement of Amino Acids of Various Subjects and the Quantity of Egg and Milk Proteins Required to Provide that

Amino Acid	*Infant (mg/kg)*	*Children (mg/kg)*	*Adult (mg/kg)*	
			Female	*Male*
Histidine	34	–	–	–
Isoleucine	119	30	8	10
Leucine	150	45	11	16
Lysine	103	60	9	11
Methionine	45	27	6	16, 3
Cystine	–	–	4	12
Total sulphur amino acids	–	–	10	16, 15
Phenylalanine	90	27	4	16, 4
Tyrosine	–	–	16	16
Total aromatic amino acids	–	–	20	16, 20
Threonine	87	35	6	7
Tryptophan	22	9	3	4
Valine	105	33	12	11
Protein Required to meet Amino Acid Need	*(g/kg)*	*(g/kg)*	*(g/kg)*	*(g/kg)*
Human milk	1.6	–	–	–
Cow's milk	2.0	0.9	0.28	0.43
Egg	1.6	0.9	0.18	0.26

The above methods have been briefly discussed in the following:

(i) Nitrogen Balance

Nitrogen is the key ingredient of protein and hence, this method is largely followed to determine the protein requirement in the body with much greater accuracy. Further, the method expresses the value in terms of *quantity of nitrogen* and thereby, the requirement is known as the exact amount of nitrogen required. Recommendations of protein by the World Heath Organization are mainly based on this method.

The method considers the actual content of nitrogen present in a protein sample and is based on the fact that most of the proteins in our regular diet contain about 16 per cent nitrogen and thus, 1 gram of nitrogen could be considered *equivalent to* 6.25 gram of protein. In other words, 1 gram of nitrogen could be obtained by the body from 6.25 gram of protein taken in the diet. The dietary protein is accordingly determined in respect of amount of nitrogen which is present in a particular food.

Loss of nitrogen from the body takes place regularly through urine, faeces and from the skin (= loss by sweat and the integument). Now, if the amount of loss of nitrogen from the body (= output of nitrogen) is found to be same with the amount that is consumed (= intake of nitrogen) it indicates that no growth takes place in the body and also that, no protein is stored in the body. Such a state of the body is referred to as *nitrogen equilibrium* in the body of the respective person. But if the consumption of nitrogen, *i.e.*, intake is greater than the loss, the body is considered to be in a state of *positive nitrogen balance*. Such condition occurs in the body of person when their muscles are developing, *e.g.*, in growing children and adolescents, patients who are convalescing after illness, pregnant women, athletes *etc*. On the other hand, if the intake of nitrogen is lesser than the loss of it, the condition is referred to as *negative nitrogen balance*. In starvation, underfeeding and wasting disease, negative balance is met with in persons. Evidently, for maintenance of nitrogen equilibrium, *minimum requirement of protein* becomes necessary. The minimum requirement of protein has been considered by many as 7 gram of nitrogen, *i.e.*, 44 gram of protein but it is not considered so by other experts.

(ii) Obligatory Loss of Nitrogen

Swaminathan (1974) in his book has provided a consolidated chart on the endogenous loss of nitrogen in the human beings as put forward by nutritional expert groups of the Food and Agricultural Organization of the World Health Organization, Food and Nutrition Board NRC, U.S.A., expert panel of the U.K., Nutrition Expert Group of the Indian Council of Medical Research and by Swaminathan (1971). The reports of the FAO/WHO, I.C.M.R. and Swaminathan have been presented in Table 8.

Table 8. Estimates of Losses of Nitrogen in Human Beings

Author	*Age-Group*	*Urine Nitrogen (mg/basal calorie)*	*Skin (Sweat + integumental loss)*	*Faeces*	*Total Nitrogen (mg/basal calorie)*
FAO/WHO Expert Group	All age groups	2.0	20 mg/kg body weight	20 mg/kg body weight	-
Recalculated per basal calorie	Child (2-3 yrs)	2.0	0.4 mg/basal calorie	0.4 mg/basal calorie	2.8
ICMR Nutrition Expert Group	All age groups	1.5	20 mg/kg body weight	20 mg/kg body weight	-
Recalculated per basal calorie	Child (2-3 yrs)	1.5	0.4 mg/basal calorie	0.4 mg/basal calorie	2.3
	Adult	1.5	0.8 mg/basal calorie	0.8 mg/basal calorie	3.1
Swaminathan (1971)	All age groups	2.0 (Urine + Sweat per basal calorie	0.1 mg/ basal calorie (integu-mental losses	0.6 mg/basal calorie	2.7

(iii) Factorial Method

This method as put forward by different groups of nutritional experts applies the following formula.

R = U + F + S + G

Where R = requirement of nitrogen, U = loss of endogenous nitrogen in urine, F = loss of endogenous nitrogen in faeces, S = loss of nitrogen through skin, *i.e.*, sweat and integument and G = nitrogen which is required to sustain growth. The factorial method was although used earlier but not much in vogue at present.

According to the World Health Organization of FAO (1965), the requirement of protein is best expressed as *reference protein value* of Net Protein Utilization (NPU) of 100. The amino acid profile of egg protein is to be considered as 100 and an additional allowance of 6 gram and 15 gram of the reference protein should be given respectively during pregnancy and lactation to a woman.

The nutritional expert groups all over the world suggest that *quality of protein* is the prime factor in determining requirement and recommendation of protein. This implies that, if the protein is of high standard, *i.e.*, of high biological value, the quantity of it that is to be recommended might be less. *Egg protein is considered to be of best quality and is assigned a score mark of 100 out of 100.* In consideration to that, if recommendation of protein for an adult is 0.7 gram per kilogram of body weight in terms of egg protein, it should be at least 1.0 gram if protein, is derived from mixed vegetables.

The booklet entitled, "Nutrient Requirements and Recommended Dietary Allowances for Indians", published by the expert group of the Indian Council of Medical Research (2004) states that keeping in view the various reports that have appeared and the recommendations given by the Food and Agricultural Organization in 1985, the nutritional expert group of the Indian Council of Medical Research has adopted the following principles in regard to fixing requirement of proteins for the Indian people.

(i) Nitrogen balance both for short-term and long-term should form the basis to determine minimum protein requirement.

(ii) Daily requirement of protein needs to be expressed in terms of gram per kilogram of body weight.

(iii) Requirement should be determined according to age of the subjects and on multiplication with standard body weight with respect to age.

(iv) The co-efficient of variation of 12.5 per cent that had been considered by the FAO committee in 1985 should be used for the subjects of all age-groups.

(v) Nitrogen values of 8 and 10 milligram per kilogram of body weight respectively for adults and children as suggested by the 1985 FAO expert committee are accepted for endogenous loss of proteins.

(vi) The body weight standards have to be considered.

(vii Provision of protein as milk protein amounting to 0.5 gram per kilogram of body weight is adequate and for other age-groups, minimum protein as suggested by FAO are considered.

Recommendations

The recommended dietary allowance of protein for the Indian people as prescribed by the National Institute of Nutrition, Hyderabad under the Indian Council of Medical Research in 1989 is as given in Table 9.

Table 9. Dietary Allowance of Protein for Indian People as Recommended by National Institute (Indian Council of Medical Research), Hyderabad

Group	*Particulars*	*Body Wt. (kg)*	*Protein (gram per day)*
Men	Sedentary work	60	60
	Moderate work	60	60
	Heavy work	60	60
Women	Sedentary work	50	50
	Moderate work	50	50
	Heavy work	50	50
	Pregnancy	50	+ 15
	Lactation –		
	0 – 6 months	50	+ 25
	6 – 12months	50	+ 18
Infants	0 – 6 months	5.4	2.05/kg
	6 – 12 months	8.6	1.65/kg
Children	1 – 3 years	12.2	22
	4 – 6 years	19	30
	7 – 9 years	26.9	41
Boys	10 – 12 years	35.4	54
Girls	10 – 12 years	31.5	57
Boys	13 – 15 years	47.8	70
Girls	13 – 15 years	46.7	65
Boys	16 – 18 years	57.1	78
Girls	16 – 18 years	49.9	63

It needs pointing out that requirement of protein to the children should be critically reviewed as a recent report has brought it to light that in India, 43 per cent of the children below 5 years are underweight, 48 per cent have stunted growth and 75 per cent of them are anaemic. The report has also stated that upto six years are the formative years in a child's life as they gain 60 per cent of the adult height, 30 per cent of the adult weight and 90 per cent of brain development. Hence, there should not be any deficiency of protein or any of the nutrient of this age.

3.14. Mutual Supplementation of Amino Acids

Proteins derived from plant-source foods have a great importance in India where considerable part of the population live on vegetarian diet. There is however, a drawback in the plant-source foods in that, the proteins in them are deficient in one or more amino acids, in other words, they comprise as *second-class proteins*. Again, it is also found that although in a particular food-staff, one or more amino acids are deficient, it is not so in another food-stuff. Thus, if cooking of food is done by making a combination of these two types of foods, deficiency of amino acid in one will be suitably compensated by the other and *vice versa*. For example, it is observed that protein of rice, wheat and other cereals is deficient of the amino acid, *lysine* but they are good sources of the amino acid, *methionine*. On the other hand, pulses are rich in lysine but they are deficient in methionine. Hence, if the food is prepared by making a mixture of cereal and pulse, deficiency of one or more amino acid, if any in the cereal may be compensated by the pulse and *vice versa*.

It may be stated in this context that in India, it is an agelong practice to take diet by mixing 5 parts of cereal (say, rice) with 1 part of pulse (= daal) and hence, in diet of such a type, there will be no deficiency of either lysine or methionine. Such type of diet which is made by judicious combination would ensure *mutual supplementation of the amino acids*. Supplementation may be done in various ways, *e.g.*, soybean which is good source of the amino acid, lysine but poor source of methionine with sesame which is poor source of lysine but rich source of methionine and in this way, many other combinations could be made.

Mutual supplementation of amino acids may be done by making a combination of plant-source food and animal-source food as well. Thus, improvement of amino acid status may be brought about when mixture is done with rice and fish, maize and fish, wheat and milk and so on. Protein efficiency ratio of the food goes high when such combined preparations are done however, if the combination is done intelligently.

It may also be stated that natural combination may take place in the same food. Wheat may be cited as an example. There are two types of protein in wheat grains, which are gliadin and glutenin. The amino acid, lysine is deficient in gliadin but glutenin is an abundant source of this amino acid. When flour is made from the wheat grains, it contains both gliadin and glutenin in suitable proportion and thus, there is no deficiency of any of these amino acids in the flour which is the edible material.

3.15. Merits and Demerits of Proteins of Plant and Animal-Source Food

Dietary proteins are obtained from both plant-source and animal-source foods. Although proteins from animal-source foods are of higher biological value than plant-source (proteinaceous) foods but the latter have a great significance in a country like India. In India, 29–30 per cent of the population do not take meat, fish or egg. In some States of India, the percentage is even much higher. For example, in Rajasthan, 73.2 and 76.6 per cent, in Haryana, 68.5 and 70 per cent and in Punjab, 65.5 and 68 per cent, respectively of the male and female population are at present, found to be vegetarians. However, a part of them take milk and hence, they should rather be considered as *lacto-vegetarians* and not strictly vegetarians. The remaining part lives exclusively on plant-origin foods. In the world also, a section of people numbering 375 millions live only on plant-origin foods. Accordingly, to meet the requirement of protein, people who are strict on vegetarian food habit have to depend on foods derived from plants.

It may be stated that the proteins whether obtained from plants or from animals have both their merits and demerits. Some of these have been stated in the following:

(a) Merits and Demerits of Plant-Source Proteins

The plant-source food-stuffs which contain high amount of protein also contain high amount of *dietary-fibres*. Hence, consumption of those foods ensures intake of dietary fibres in their body. Vegetarian foods contain *anti-oxidants* and *low amount of saturated fat*, for which they are beneficial to cardiovascular health. The risk of *type-2 diabetes* and increase of cholesterol and fat in the body is also much less with these foods. Besides, vitamins and minerals content of the protein-rich, plant-source foods are higher than those of the animal-source foods. Mention may be made of pulses which are rich sources of protein contain *saponins* that tend to reduce cholesterol level, improve immunity in the body and protect against cancer. The plant protein, *lectin* found in wheat, rice, pea, beans, lentil, soybean *etc.* has cancer preventing property.

The proteins derived from plant-source foods have however, the greatest drawback in that, their quality, *i.e.*, *biologocal value* is low as compared to the proteins derived from animal-source foods, which indicates that their amino acid composition is poor. But the amino acid deficiency of many plant-source foods could be overcome if foods of two or more types are *mixed together and taken.* It may be mentioned that the type of proteins that are found in many cereals and pulses mostly come to the category of simple protein, *e.g.*, albumins, globulins, glutelins, prolamines, protamines *etc.*

The proteins, *viz*, gluten and gliadin present in wheat and some other cereals are intolerant to some people. A disease known as *caeliac sprue* appears in them when these proteins go inside their body, which is an intestinal disease that appears in them. This is characterized by malabsorption, loss of weight, distention, bloating, diarrhoea and steatorrhoea. Diagnosis of the disease, which is also known as tropical sprue is difficult and diet free from these proteins is the only remedy. Many of the pulses are known to contain some toxic substances. Examples may be cited as follows: (i) Lathyrism is a health hazard which is characterized by muscular weakness and paraplegia and the neurotoxin BOAA present in chickling pea (= khesari daal) is responsible for this. (ii) Favism is another allergic reaction which causes fever, vomiting, diarrhoea and haemolytic anaemia in people who have an inherited deficiency of the enzyme, glucose – 6 – phosphate dehydrogenase. The fava beans (= broad beans) and even pollen grains of the flowers of these plants cause this abnormality. (iii) Coumarins present in some beans cause damage to kidney and liver. (iv) Tannins, the phenolic compounds present in some pulses may cause vomiting, stomach irritation and liver trouble. (v) Goitrogens present in soybean and groundnut may cause goitre. (vi) Aflatoxicosis is a serious poisoning caused by the intake of groundnuts, if these are contaminated by the species of the *Aspergillus* fungus which contains aflatoxin.

Suitable processing of the pulse grains may however, remove out many of the toxins. Fox example, (i) steeping the grains overnight after dehusking and then steaming for half an hour remove the toxin, BOAA from chickling pea. (ii) Soaking and parboiling destroy trypsin enzyme inhibitor from soybean. (iii) Soaking and boiling in water remove cellulose from pulses to a great extent.

In the chicking pea (= knesari), an agronomic variety, *viz.*, 'P–24' has been developed which has very low content of the toxin, BOAA. It is found that as against 0.55 per cent of BOAA which is usually present in most varieties of chickling pea, the P–24 variety contains 0.3–0.5 per cent of BOAA in the grains. Low content of protein is a drawback in the plant source foods. But it may be mentioned that somc agronomic varieties of cereals and pulses have been produced in which protein content has been markedly increased. Wheat may be cited as an example, the protein content in the grains of which has increased upto 14 per cent in some varieties, *viz.*, UP – 262, HD – 2327, WH – 147 *etc.* by suitable breeding procedure.

It may be brought to light that by the act of efficient breeding, many varieties of agricultural crops have been produced, protein content of which has gone to a very high level. For example, in maize, by introducing the genes, *viz.*, *Opaque – 2* and *Opaque – 7* in some varieties, the amino acid composition had been significantly improved. By such breeding procedure, the biological value of the composites, *viz.*, *Protina*, *Ratan* and *Shakti* have almost become comparable to the milk protein, casein. Similarly, in barley, the *hilly*, *Hiproly*, *Riso – 1508*, *Notch* – and *Notch – 2*, genes have conferred high lysine content in the varieties. In the variety of rice, *viz.*, CRPH – 1, the protein content has increased to 13.5 per cent.

(b) Merits and Demerits of Animal-Source Proteins

The greatest merit of animal-source protein lies in their quality. Thus, the Protein Efficiency Ratio (PER), Biological Value (BV), Net Protein Utilization (NPU) and Chemical Score of meat, fish, egg and milk are found to range from 3 to 4.5, 84 to 96, 72 to 91 and 65 to 70 respectively, while they range from 1.6 to 2.5, 58 to 67, 45 to 57 and 40 to 60 respectively in the plant-source foods like cereals, pulses and oilseeds. It may be cited that the important amino acid, lysine, deficiency of which may cause *pellagra* disease is found to range from 0.44 to 0.57 gram per gram of nitrogen in fish, meat, egg and milk but in cereals and millets, lysine content is only 0.11 to 0.23 gram per gram of nitrogen.

The sulphur containing amino acids (cystein, methionine and cystine) are also much higher in animal-source food. For example, the sulphur-containing amino acids content in animal-source foods ranges from 0.09 to 0.19 gram per gram of nitrogen but these do not exceed 0.14 gram per gram in cereals and pulses. The animal-source foods also provide the fat-soluble vitamins in the body. Bioavailability of minerals like iron and zinc and some vitamins of the B-group is again higher in these food.

Among demerits of animal-source foods, the present-day physiologists agree that at least mutton, beef, pork and egg are potent agents that raise the cholesterol and triglyceride levels in the blood besides that, these foods are devoid of dietary fibres and anti-oxidants. Accordingly, these foods are considered as harmful to heart health. Recent studies indicate that too much consumption of red meat is one of the causes of colon cancer.

For the above reasons, a great many people all over the world are switching over exclusively to vegetarian food habit. They are known to get full satiety and are maintaining good health. Prof. A. B. Singha Mahapatra in his book (2005), "Essentials of Medical Physiology" (Published by Current Book International, Kolkata) has rightly stated, "Very often the superiority of animal proteins is advocated over the vegetable proteins but that is *psychological* or for their *palatability*. It should also be noted that the vegetarians maintain good health without animal protein. "

In conclusion, it may be said that as merits and demerits are there both in vegetarian and non-vegetarian foods, a healthy individual unless otherwise advised, or on religious ground should have both vegetarian as well as non-vegetarian foods in their regular diet and also in suitable proportion.

3.16. Adverse Effects

Despite the fact that protein is considered by many as most important component of nutrition in the body, its improper intake may have adverse effect under certain conditions. Undesirable effects are produced when consumption of protein becomes *excess* and also when it is consumed from some particular food-stuffs. The symptoms of excess intake of protein in the body are constipation, high abdominal acidity, headache and excess intake for long time may cause loss of calcium through urine, body ache, gout and impairment of kidney. High consumption of protein for prolonged period may also bring about atherosclerosis and even cancer of colon, breast and prostate.

Too much storage of protein in the body on consuming high amount of whey has been reported to cause nausea, cramp, fatigue and reduced appetite. Intake of high amount of red meat and egg is known to make *great harm to the kidneys* and sometimes an unpleasant odour (like that of carnivores) may come out from the body of persons who have the habit of taking too much of red meat.

Regular intake of protein supplements in the form of powder, granules, biscuits, shakes *etc.* is a practice among many. Such direct ingestion of protein without medical advice may result in excess accumulation of protein in the body. This may hamper absorption of other nutrients in the body and result in renal, cardiovascular or other abnormalities. Direct intake of protein is sometimes advised for children, athletes and convalescing patients but the quantity to be taken and composition of the product are needed to be approved by the health care providers before using them. The amino acid, phenylalanine is nowadays used in making the artificial sweetener, aspartame, which may cause damage to the nerves if taken regularly. Topical (local) application of amino acid formulation on the scalp to stimulate hair growth may have dangerous effect if used without approval by expert trichologists or dermatologists.

3.17. Protein Content of some Food-stuffs

Protein content of some food-stuffs as listed by Gopalan *et al.* in their book, "Nutritive Value of Indian Foods," (2018) (Revised by Narasinga Rao *et al.*) and published by the National Institute of Nutrition, Hyderabad, under the Indian Council of Medical Research, New Delhi has been presented in Table 10. The amount of protein present in some selected food-stuffs have only been presented. For the amount in other food-stuffs, the original little may be consulted.

Table 10. Protein Content of some Food-stuffs under different Groups (Figures are in gram per 100 gram of edible portion)

Cereal grains			
Bajra	11.6	Barley	11.5
Italian millet	12.3	Jowar	10.4
Maize (dry)	11.1	China dhan	12.5
Ragi (marua)	7.3	Rice (parboiled, handpounded)	8.5
Rice (parboiled, milled)	6.4	Rice (raw, hand pounded)	7.5
Rice (raw, milled)	6.8	Rice (bran)	13.5
Rice (flakes, chira)	6.6	Rice (puffed, muri)	7.5
Samai	7.7	Wheat, bulgar (parboiled)	8.2
Wheat (whole)	11.8	Wheat, flour (whole)	12.1
Wheat, flour (refined)	11.0	Wheat (semolina, suji)	10.4
Wheat bread (white)	7.8		

Pulses			
Bengal gram (whole)	17.1	Bengal gram (daal)	20.8
Black gram (daal)	24.0	Cowpea	24.1
Field bean (dry)	24.9	Green gram (whole)	24.0
Green gram (daal)	24.5	Horse gram (whole)	22.0
Khesari (daal)	28.2	Lentil (masur)	25.1
Moth beans	23.6	Pea (green)	7.2
Pea (dry)	19.7	Rajmah	22.9
Red gram (daal)	22.3	Soybean	43.2
Leafy vegetables			
Amaranth	3.0	Bathua	3.7
Beet green	3.4	Brussels sprout	4.7
Cabbage	1.8	Colocasia (leaves)	6.8
Curry (leaves)	6.1	Drumstick (leaves)	6.7
Fenugreek (leaves)	4.4	Lettuce	2.1
Parsley	5.9	Spinach	2.0
Roots and tubers			
Arrow root (flour)	0.2	Beet root	1.7
Carrot	0,9	Colocasia	3.0
Dioscorea	1.3	Onion (big)	1.2
Onion (small)	1.8	Radish (pink)	0.6
Sweet potato	1.2	Tapioca	0.7
Yam (wild)	2.5	Ash gourd	0.4
Bitter gourd (small)	2.1	Bitter gourd (large)	1.6
Brinjal	1.4	Cauliflower	2.6
Cucumber	0.4	Drumstick (pod)	2.3
Drumstick (flower)	3.6	Giant chilli	1.3
Jackfruit (unripe)	2.6	Knol khol	1.1
Okra	1.9	Leek	1.8
Papaya (unripe)	0.7	Plantain (fruit)	1.4
Plantain (flower, mocha)	1.7	Plantain (stem, thor)	0.5
Pumpkin	1.4	Ridged groud	0.5
Snake gourd	0.5	Tinda	1.4
Tomato	1.9		
Nuts and oilseeds			
Almond	20.8	Cashew (karnel)	21.2
Coconut (fresh)	4.5	Gingelly seed	18.3
Groundnut	25.3	Mustard (seed)	20.0
Pistachio (pesta, kernal)	19.8	Walnut	15.6

Condiments and spices			
Asafoetida	4.0	Cardamom	10.2
Cloves	5.2	Coriander	14.1
Cumin (jira)	18.7	Fenugreek (seed)	26.2
Garlic	6.3	Mace	6.5
Nutmeg (fruit)	7.5	Jowan	17.1
Poppy (seed)	21.7	Turmeric	6.3
Fruits			
Emblica	0.5	Apricot (fresh)	1.0
Avocado	1.7	Wood apple (bael)	1.8
Banana (ripe)	1.2	Bullock's heart	1.4
Cape-gooseberry	1.8	Cherry (red)	1.1
Currants (black)	2.7	Date (dried)	2.5
Fig	1.3	Grape	0.6
Grapefruit	1.0	Guava	0.9
Jackfruit	1.9	Pithecellobium (bilati tentul)	2.7
Litchi	1.1	Lime (sour lime)	1.5
Mulberry	1.1	Orange (mandarin)	0.7
Palmyra (ripe pulp)	0.7	Peach	1.2
Perimmon	0.7	Phalsa	1.3
Plum	0.7	Pomegranate	1.6
Raisin	1.8	Rasberry	1.0
Custard apple	1.6	Elephant apple (kath bel)	7.1
Fishes and sea food			
Banspata	18.2	Bata (small)	14.3
Bhangan bata	19.4	Bhekti	14.9
Bhola	15.2	Boal	15.4
Cat fish	21.4	Chital	18.6
Crab (small)	11.2	Folui	19.0
Herring	20.3	Hilsa (Ilish)	53.7
Kalabasu	14.7	Katla	19.5
Khoyra	18.0	Koi	14.8
Lata	19.4	Lobster	20.5
Mackerel	18.9	Magur	78.5
Mahasol	25.2	Mrigel	19.5
Pabda	19.2	Parsey	17.6
Pomfret (black)	20.3	Singhi	22.8

Meat and poultry			
Beef (muscle)	22.6	Duck	21.6
Egg (duck)	13.5	Egg (hen)	13.3
Fowl	25.9	Goat meat	21.4
Mutton (muscle)	18.5	Pigeon	23.3
Pork (muscle)	18.7	Snail, small	12.6
Venison	21.0		
Milk and milk products			
Ass's milk	2.1	Buffalo's milk	4.3
Cow's milk	3.2	Goat's milk	3.3
Human milk	1.1	Curd of cow's milk	3.1
Channa (from cow's milk)	18.3	Cheese	25.1
Sugars			
Sugarcane	0.4	Honey	0.3
Jaggery (gur from cane juice)	0.4	Jaggery (gur from coconut palm juice)	1.0
Jaggery (gur from date palm juice)	1.5	Jaggery (gur from fan palm juice)	1.0
Beverages (Non-alcoholic)			
Neera	0.4	Sugarcane juice	0.1
Beverages (Alcoholic)			
Pachwai (Assam)	3.0	Toddy (fermented)	0.1
Toddy (sweet)	0.1		

Chapter 4

Fats

Fat is one of the most important nutritional constituents in the body of human beings. Out of the six components of nutrients, it ranks next to protein. This item of food is highly palatable and is valued as the *most efficient and highly concentrated form of energy* in the body. Being insoluble in water, it is not easily lost from the body and thus, is stored up in the body for considerable period.

Fat constitutes about 12 per cent of the total body weight in human beings and it is rich in the nervous tissue. In other parts of the body also, it is abundant or high. For example, it is present to about 50 per cent in subcutaneous tissue, 20 per cent in mesentric tissue, 15 per cent in peritoneum and so on. In the cell, much fat exists both in the cytoplasm as droplets or as emulsion and also in the cell-membrane where it provides structural support.

Camels and desert animals store considerable amount of fat in their body in order to get water in the metabolic form (on breakdown of fat) so as to quench their thirst where water source is not available. (Some people have the notion that the camels store water in the hump on their back. This is not correct and camels do not store water. Instead, they store fat in the hump and the fat metabolically converts to liberate water which they use in thirst).

Fats are of great importance in the physiology of plants also. Some plants store fats in the body for later use. The higher plants store large amount of fat in their seeds and nuts. Brazil nut and coconut are known to contain more than 65 per cent fat in their kernels (= seeds). Considerable amount of fat is stored by many cereal and pulse grains and oilseeds and these are particularly rich sources.

At the present time, many physicians and dietitians advise their patients to restrict consumption of fats, particularly of certain types of fats for their harmful effects on cardiovascular system, functioning of brain and getting rid of abnormalities like obesity, joint pain and even cancer. However, it may also be pointed out that fats provide plenty of other benefits in the body and some of these have been stated later under the heading on functions and benefits of fat. As a matter of fact, whatever food we take, whether those are of plant or of animal origin, we cannot completely eliminate fat. Judicious nutritionists advise that 20 to 35 per cent of our daily intake of food should actually come from fatty substances and drastic reduction to that might effectuate serious health hazards and may even lead to death.

4.1. Definition

To a common man, fat and oil are conceived in a different way. Fat is considered to be a food when it is solid or semi-solid in texture, not soluble in water and on touching, though the fingers become greasy they do not very easily dip into it. On contrary, oil is absolutely liquid, not soluble in water but unlike fat, the finger easily dips in it like water.

In a general usage, the solid fat is termed as *visible fat* while the oil is called *invisible fat*. The solid form of fat (= visible fat) on heating, converts to oil. But it reverts to the solid form of fat when chilled. In fact, the fat and oil are same and the terms are used interchangeably.

It should however, be remembered that the term, fat which has derived from the Anglosaxon word, *faett* includes both fats and oils. In chemical terms, all fats and all organic forms of oil belong to the same group and the group is termed *lipid*. Thus, a lipid matter includes both fat and oil. Besides these, some other compounds related to fats and oils also come under the broad group of lipids. Chemically, these materials are formed due to reaction of *fatty acids* and *glycerol* (= glycerine) and hence, all lipid materials are formed as *esters* of fatty acids and glycerol, *i.e.*, fatty acid glycerides.

4.2. Functions and Benefits

Fats have plentiful functions in the body and the more important ones from physiological standpoint may be stated as follows:

(i) Most important function lies in the fact that they *provide energy*. Unlike carbohydrate and protein, fats store energy in very high amount. While one gram of carbohydrate or protein provides 17.57 kilojoules of energy in the body on consumption, one gram of fat yields 37.8 kilojoules of energy. Thus, fat stores energy in a *highly concentrated form*.

(ii) By providing energy, fats and oils save the proteins in regard to release of energy by degradation. That is, fats act as *protein sparers* like carbohydrates. This has been stated in the Chapter on Carbohydrates (Heading 2.3).

(iii) Fats are existent in the cell-membranes and due to that, they help in maintaining *structural integrity of cells.*

(iv) The vitamins A, D, E and K are fat soluble and not soluble in water. Thus, these vitamins are *made available to the body* through the fat or oil medium only.

(v) Presence of fatty substance under the skin acts in *insulating* the body against heat and cold. Penguins and other animals which inhabit in extremely cold climate deposit very high fat under their skin for this purpose.

(vi) Essential fatty acids, *i.e.*, linoleic acid, linolenic acid and arachidonic acid greatly help the body to maintain *growth, skin health, reproduction* and particularly to *prevent atherosclerosis.*

(vii) Fatty acids which are obtained by degrading fats or oils are important sources in formation of many compounds.

(viii) Desert animals like camels and kangaroo-rats store excess fat in the body to procure metabolic water in thirst.

(ix) Hibernating animals store much fat in the body in winter to get metabolic water and energy in hibernation.

(x) Migratory birds store high fat in body before migration which metabolize during their long flight and thus, provides food, energy and water to quench thirst.

4.3. Properties

Important physical properties of fatty substances may be stated as follows:

(i) Texturally, they are solid, semi-solid or liquid.

(ii) They are felt greasy when touched.

(iii) Most of the fats and oils in their pure form have no colour or odour.

(iv) All fats and oils float in water and thus, indicate their low specific gravity.

(v) Temperature which is required to melt differs according to type of fats.

(vi) When poured on water, fats and oils spread on the water surface *uniformly* even if very small amount is poured.

(vii) Although fat is not soluble in water, if fat and water are mixed and vigorously shaken in a container, fat is broken down into very fine droplets and remain uniformly dispersed in water. This is called *emulsion*, such as milk is an emulsion of fat globules in water.

Important chemical properties of fatty substances may be enlisted as follows:

(i) They are soluble in some *organic solvents, e.g.*, ether, acetone, chloroform, alcohols *etc.* and not in water.

(ii) In pure form, most of them have neutral or very close to *neutral reaction.*

(iii) *Hydrolysis* – when reacted with acids, alkalis, highly heated steam or the enzyme lipase, fatty molecules are broken down and the parent materials, *i.e., glycerol and fatty acids are released.*

(iv) *Rancidity* – Natural fats which are not treated with chemicals tend to oxidize spontaneously. As a result, the fats emit bad smell and become unfit for consumption. This is called *rancidity of fats*. Rancidity may be of three types, which are *oxidative* in which some metals like copper hasten oxidation, *hydrolytic*, which is done by lipase enzyme and *ketonic*, which is done by some fungi and notably by the fungal species, *Aspergillus niger* and *Penicillium glaucum*. Rancidity of fat is determined by examining peroxidase value, carbonyl compounds or by thiobarbituric acid of the fat. Rancidity could however, be much controlled if the fat is treated with anti-oxidant chemicals, *e.g.*, vitamins E, C, gallic acid, carotene *etc.* before it is stored and the container is tightly stoppered. Transparent containers should also be avoided.

(v) *Saponification value* – In soap manufacturing industries, fats are hydrolyzed by treating with alkalis. This is called *saponification*. Thus, for a given fat sample, the number of milligrams of the alkali, *i.e.*, potassium hydroxide (KOH) which is required to saponify one gram of the fat sample gives the saponification value of that sample.

(vi) *Iodine value* – This is another index which accounts for the amount of unsaturated fatty acids that are present in a given fat sample. This is determined in consideration to the gram of iodine that could be absorbed by 100 gram of the given fat sample.

(vii) *Reichert-Meissl value* – This is a measure which is determined in consideration to the number of millilitres of deci-normal solution of an alkali (NaOH or KOH) that are required to neutralize fatty acids (steam volatile) which exist in a given fat sample weighing 5 gram.

(viii *Thiocyanogen value* – The amount of thiocyanogen that could be assimilated by 100 gram of a given fat sample is considered to determine this value of the fat sample.

4.4. Metabolism of Fat

4.4.1. Synthesis (= Anabolism)

Chemically, fat is a *triglyceride of fatty acids*. That is, *one molecule* of glycerol (= glycerine) is esterified (= condensed) with *three molecules* of fatty acids and this gives rise to one molecule of fat.

To have an understanding of fat metabolism, knowledge on glycerol and fatty acids is necessary. Hence, the subject (= anabolism) is discussed under three parts : (1) Glycerol, (2) Fatty acids and (3) Formation of the triglyceride ester (= fat).

1. Glycerol

This is actually an *alcohol.* Alcohols are hydroxy derivatives of alkanes. They are classified according to the number of hydroxyl groups (– OH) that are present in them. That is, when the alcohols contain 1 hydroxyl group, it is termed monohydric alcohol, when contain 2 hydroxyl groups, they are termed as dihydric alcohols, when contain 3, they are termed as trihydric alcohols and so on. Glycerol is a *trihydric alcohol,* which means its molecule contains 3 hydroxyl groups as shown below:

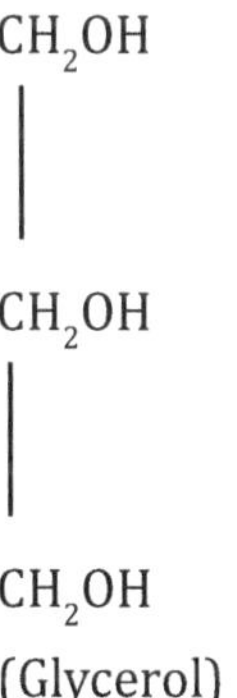

(Glycerol)

Glycerol (IUPAC name : Propane-1, 2, 3-triol) which is known in industries as *glycerine* is a heavy liquid, which is colourless, odourless, hygroscopic and is soluble in water and ethanol (= ethyl alcohol).

Glycerol is synthesized from the hexose sugar, *glucose.* From glucose, two molecules of triose-phosphate are formed, which are dihydroxy-acetone phosphate and glyceraldehyde-3-phosphate. Glycerol is formed from the latter.

2. Fatty Acids

Carboxylic acids are those which contain one or more carboxylic (– COOH) groups. A carboxylic acid is formed by combination of a carbonyl (– C = O) group and a hydroxyl (– OH) group. Carboxylic acids are classified as monocarboxylic acid, dicarboxylic acid, tricarboxylic acid *etc.*, according to the number of carboxyl groups that are present in their molecules.

Monocarboxylic acids of aliphatic group are usually termed as *fatty acids.* This is because, many of them and particularly those of the higher member (having 12 to 18 carbon atoms), *e.g.*, palmitic acid, stearic acid *etc. constitute fats.* These acids are divided into *saturated* or *unsaturated* group in accordance with nature of the carbon chain which gets attached to the carboxylic group as shown below:

CH_3CH_2COOH

(Saturated monocarboxylic acid)

$CH_2 = CHCOOH$

(Unsaturated monocarboxylic acid)

Some of the common fatty acids belonging to the saturated and unsaturated groups are stated in the following tables.

Table 11. Saturated Fatty Acids

Common Name	*Systematic Name*
n-Butyric	–
Caproic	*n* - Hexanoic
Caprylic	*n* - Octanoic
Capric	*n* - Decanoic
Lauric	*n* - Dodecanoic
Myristic	*n* - Tetradecanoic
Palmitic	*n* - Hexadecanoic
Stearic	*n* - Octadecanoic
Arachidic	- Eicosanoic

Table 12. Unsaturated Fatty Acids

Common Name	*Systematic Name*
Palmiloleic	9 – Hexadecenoic
Oleic	*cis* – 9-Octadecenoic
Elaidic	*trans* – 9 - Octadecenoic
Vaccenic	11 - Octadenoic
Linoleic	*cis* - 9, 12 - Octadecadienoic
Linolenic	9, 12, 15 - Octadecatrienoic
gamma –"	6, 9, 12 - Octadecatrienoic
Eleostearic	9, 11, 13 - Octadecatrienoic
Arachidonic	5, 8, 11, 14 – Eicosatetraenoic

Essential Fatty Acids

In connection with fatty acids, it becomes necessary to have knowledge on *essential fatty acids* for their high nutritional importance. Burr and Burr in the late 1920s established their beneficial role on experimenting with white rats.

Essential fatty acids are : *linoleic acid, linolenic acid* and *arachidonic acid.* These polyunsaturated fatty acids have a number of benefits in the body among which the following are important. (i) They retard or prevent cholesterol accumulation in the blood and thereby, save the blood vessels against *atherosclerosis.* (ii) Importance of them is known for growth of children. (iii) Deficiency of them causes skin lesions and eczema, particularly in children. (iv) They have much benefit in pregnancy and lactation in women. (v) Metabolites of arachidonic acid (eicosanoids) are of great importance.

Oils that are obtained from plant sources are rich in these acids while fats or oils which are derived from animals are poor sources. Essential fatty acids are required in minute amount only but it should be noted that even that amount is of absolute necessity.

Fatty acids are synthesized from *acetate* which gives rise to acetyl CoA on activation by ATP. Conversion of PSCR takes place by hydrolysis of phospho-enol-pyruvic acid into pyruvic acid and pyruvic acid reacts with CoA to give rise to acetyl CoA. Splitting of TPP-glyceraldehyde additive compound with thiamine may also take place. Acetyl phosphate compound which is formed may be translated with CoA to form again inorganic phosphate and acetyl CoA. When acetyl CoA is formed, higher fatty acids may be formed by ATP and NADH. Fatty acids may also be formed due to hydrolysis of fat by lipase enzyme. Unsaturated fatty acids are formed from saturated fatty acids.

3. Esterification (= condensation) of Fatty Acids and Glycerol

In the first step of triglyceride formation, the fatty acids combine with CoA. To form one molecule of triglyceride, three fatty acids are reacted with three CoA molecules. In the second step, the three fatty acids-CoA complex react with one molecule of glycerol. Triglyceride molecule, *i.e.*, fat is thus, formed as shown in the following figure and eventually, the three CoA molecules are released.

$$\begin{matrix} CH_2OH \\ | \\ CH_2OH \\ | \\ CH_2OH \end{matrix} + \begin{matrix} C_{15}H_{31}COOH \\ C_{15}H_{31}COOH \\ C_{15}H_{31}COOH \end{matrix} \rightleftharpoons \begin{matrix} C_{15}H_{31}COOCH_2 \\ C_{15}H_{31}COOCH_2 \\ C_{15}H_{31}COOCH_2 \end{matrix} + 3H_2O$$

1 molecule of glycerol	3 molecules of a fatty acid (= palmitic acid)	1 molecule of a triglyceride (= tripalmitin)

Synthesis of a Triglyceride (= fat) Ester.

4.4.2. Breakdown (= Catabolism)

Breakdown, *i.e.*, catabolism of fat *releases energy.* This involves breaking down, *i.e.*, degradation of fat (= triglyceride) into glycerol and fatty acids. Degradation, *i.e.*, *oxidation of glycerol and fatty acids takes place separately.*

The process in each case is stated below:

4.4.2.1. Oxidation of Glycerol

Glycerol which is released on hydrolysis of fat changes to glycerol-phosphate on reaction with ATP. This is oxidized or dehydrogenated by phosphate dehydrogenase enzyme and is converted to the compound, 3-phospho-glyceraldehyde which enters into glycolytic pathway and is broken down to release energy.

4.4.2.2. Oxidation of Fatty Acids

(It may be pointed out that normally glucose molecules which are stored in the body in the form of glycogen molecules are oxidized to release energy which is required by the body. Glycogen remains abundantly stored in muscles and liver. Besides, it is also much stored in the brain, stomach, kidney, WBC *etc.* However, when concentration of glycogen becomes lower, fatty acids may be oxidized to release energy).

When the fatty acids are oxidized, they form acetyl CoA, *i.e.*, become activated acetate. These enter into the TCA cycle and thereby, release energy. The oxidation process of fatty acids is however, complex and several postulations have been put forward by physiologists to explain the phenomenon.

The concept of *beta-oxidation* put forward by Knoop is much accepted in regard to oxidation of fatty acids. According to this viewpoint, fatty acids are oxidized at the carbon atom in the beta position of the carboxyl group. Due to that, cleavage of the two terminal atoms takes place and is left behind a fatty acid with two carbon atoms fewer than that of the original. Fatty acids which have higher carbon atoms are actually oxidized by beta-oxiation.

The concept of *omega oxidation* of fatty acids put forward that a dicarboxylic acid of long chain is at first formed due to oxidation of terminal carbon atom to carboxyl group. Beta-oxidation then starts in it and a dicarboxylic acid is formed.

Ketosis is a health abnormality that results in excess accumulation of ketone bodies in blood, urine or other tisses. Oxidation of fatty acids by the liver results in formation of ketone bodies. When there is *high rate of accumulation of acetyl CoA* but their capacity for disposal is less, ketosis develops. This is due to impaired metabolism of carbohydrate and results particularly in diabetic condition and in starvation.

4.5. Lipid

In the foregoing pages, only fats and oils have been discussed. It should however, be noted that fats and oils come under a large group. This large group is *recognized as lipids.* In organic chemistry, lipid is in fact, a large group and within this group, fats and oils are placed.

The term, lipid, which is also called lipid has derived from the Greek word, *lipose.* This broad group contains fats and oils but apart from fats and oils, some other organic compounds which are related to those, *i.e.*, have similarity with fats and oils are also placed in this group. Many such compounds are there and the important ones may be stated as waxes, steroids, phospholipids, glycolipids, sulpholipids, lipoproteins, four vitamins, *e.g.*, vitamin A, D, E and K and the like.

Thus, the triglycerides, fats and oils come under the broad group of lipid. However, among all lipids, fats and oils are most abundant in nature. It may be mentioned in this context that the lipids are *descriptive* and not of chemical specificity like protein or carbohydrate.

4.6. Classification of Lipids

Lipids have been classified by various authors in many ways. A simple classification consists of bringing them under two broad groups.

(I) *Simple lipids* and (II) *Complex lipids.*

Simple lipids are referred to those which are relatively simple in structure, although always they are not so. Examples of simple lipids are fats, oils, some other fatty acid esters such as butter, some oils obtained from plants etc. as well as waxes and steroids which are also placed in this group.

Under the group (II), *i.e*, complex lipids, those lipids which are complex in structure are placed. Some examples of lipids which have been maintained under this group are phospholipids, lecithin, cephalin, glycolipids, sphingolipids etc. The classification of such type of grouping of lipids is however, not much accepted.

In another classification, all lipids are brought under the following four groups. These are, (I) *Simple lipids*, (II) *Compound lipids*, (III) *Waxes*, and (IV) *Derived lipids.*

Among the above four groups, simple lipids are considered to those lipids which are *esters of glycerol and fatty acids, i.e.*, fats and oils. When these esters (= triglycerides) remain in liquid form at 20°C temperature, they are called oils while those that remain in solid or semi-solid form at 20°C are recognized as fats. Under the group of compound lipids, those lipids which contain some other chemical(s), in addition to glycerol and fatty acids are placed. The lipids under the wax group are considered to those which are esters of fatty acids and long-chain alcohols. Derived lipids are those which are formed due to degradation of the lipid structures, *i.e.*, catabolized products of lipids, *e.g.*, glycerol, fatty acids, alcohols, sterols *etc.*

There are many other ways of how lipids are classified. A classification of lipids presented by L. Veerakumari (2004) in her book, *Biochemistry* (MJP Publ., Chennai-5) appears to be more explanatory. In this classification (by Bloor), all lipids have been classified into four groups as follows. (I) *Simple lipids*, (II) *Compound lipids*, (III) *Derived lipids* and (V) *Compounds associated with lipids.*

The compounds under these four groups have been stated as follows.

Group (I) : Fats, waxes and their components.

Group (II) : Phospholipids (lecithin, choline, cephalin), glycolipids, sulpholipids and lipoproteins.

Group (III) : Fatty acids and glycerol.

Group (IV) : Steroids (sterols), bile acids, sex hormones (male and female), saponin, cardiac glycosides and toad poison.

4.7. Characteristic Features of some Non-Fat Lipids

Physical and chemical properties of fats and oils have been stated earlier. Characteristic features of some other lipids which are not fat have been discussed under this heading.

4.7.1 Simple Lipids

(A) Wax

Like fats, waxes are also esters of fatty acids but they are esters of *long-chain alcohols* and not of glycerol. Waxes are not soluble in water and like fats and oils, they are soluble in fat-solvents, *e.g.*, ether, acetone, chloroform etc. However, like fats or oils, they are not hydrolyzable by lipase enzymes.

Waxes have great importance to serve as a *protective layer* of some organisms. In plants, they are of great help to control evapo-transpirational loss of water. The wax present on the feathers of birds protect them against environmental hazards.

There are different types of waxes. Most common is *bees wax*, which is chemically, myricyl palmitate $[CH_3(CH_2)_{26} \bullet CH_3]$. Lanolin is another wax of much importance. It is obtained from wool of sheep and hence, it is also called *wool fat*. This is the only lipid that *absorbs water*. Lanolin is much used as medium in making many ointments, cosmetics and plant growth regulators (= plant hormones) in horticulture and plant physiological researches. Among other waxes, *sperm-whale wax, carnauba wax* and *Chinese wax* are used in some industries.

4.7.2. Compound Lipids

Lipids of this group are also chemical esters of fatty acids but in addition to glycerol and fatty acids, they contain in them *some other chemical(s) also*. In accordance with presence of phosphate, sugar, sulphate and protein, they are recognized as (A) Phospholipids, (B) Glycolipids, (C) Sulpholipids and (D) Lipoproteins. These have been stated below :

(A) Phospholipids

These lipids are also known as *phosphatides* and contain extra chemicals as phosphoric acid with nitrogenous base. They are much present in some organs, *e.g.*, brain, kidney and heart and have a great role in *blood coagulation process and in preventing fatty liver*. Some of the lipids under this group are as follows.

(a) *Lecithin* — This lipid contains in its molecule one molecule of glycerol which is esterifed with two molecules of fatty acids. Besides this, it also contains phosphoric acid and nitrogenous base. It has a great role in fat metabolism.

(b) *Choline* — This is a constituent of lecithin and *sphingomyelin.* (Spingomyelin is made up of fatty acids, phosphoric acid, choline and a complex *amino-alcohol, sphingol or sphingosine*). Deficiency of choline in the body causes *fatty liver* and hampers transmission of nerves also. Choline has role in protein metabolism as well.

(c) *Cephalin* — It occurs along with lecithin and apart from glycerol and phosphoric acid, it contains an alcohol-amine compound (= ethanolamine).

(d) *Phosphatidyl serine* — Besides fatty acids, it contains glycerol, phosphoric acid and serine.

(e) *Plasmalogens* — It differs from lecithin and cephalin in that, one of the fatty acids in it is replaced by saturated ether.

(f) *Phosphoglycerides* — This is formed on condensation of three molecules of glycerol and two molecules of phosphoric acid.

(B) Glycolipids

These lipids always contain *sugar* in their molecules. Important lipids under this group are cerebrosides and gangliosides. Cerebroside contains glucose or galactose, fatty acid, sphingosine but do not contain glycerol or phosphoric acid. Ganglioside contains neuraminic acid in addition to fatty acid, glucose or galactose and sphingosine.

(C) Sulpholipids

Sulphuric acid is always present in the molecules of these lipids. Besides this, galactose, potassium and sphingosine are also present in them. Brain contains high amount of these lipids.

(D) Lipoproteins

Lipids of this group are rich in protein molecules. Vitamins A, D, E and K, cholesterol, licithin etc. come under this group.

4.7.3. Related Lipids

Important compounds belonging to this group are as follows:

(A) Steroids

Steroids or sterols constitute a large number of compounds which are not fat but recognized as lipids. They are aromatic, polycyclic alcohols with nucleus core as perhydro-cyclo-pentano-phenanthrene and have 17 carbon atoms. Sterols remain associated with *phospholipids*. They may remain as free or as esters, *i.e.,* sterides. Sterols are *synthesized from non-sterol source* and are soluble only in fat solvents. They occur in animals, human beings and plants. Accordingly, they are grouped as animal sterols, *e.g.,* cholesterols, plant sterols, *e.g.,* phytosterol and mycosterols, *e.g.,* ergosterol.

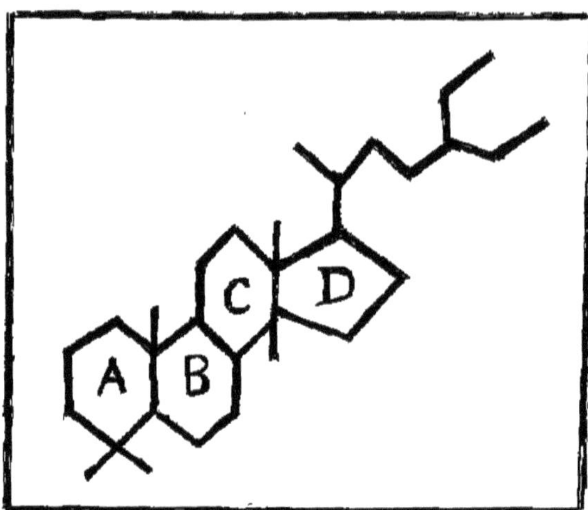

Figure 13. Steroid Ring Structure.

In the human beings and animals, sterols are present in higher amount in nervous tissue, ovary, testes, liver *etc.* They are present in all cells and both inside the cytoplasm as well as in the cell-membrane. Major sterols in the human body are cholesterol, male and female sex hormones, coprosterol, vitamin D, bile acids (cholic acid and derivatives), adrenocortical hormones such as androgens, mineralocorticoids *etc.* Characteristic features of some steroids are briefly stated in the following:

(a) Cholesterol

Cholesterol is considered to be the most important of the sterols. The name is derived from the Greek word, *chole* (= bile) and *stereos* (= solid) with *-ol* (= alcohol). The compound was first identified in solid form in gallstone by Francois Poulletier Salle in 1769. It was named as cholesterine in 1815 by Michael Eugene. Cholesterol is a monohydric, secondary alcohol.

(i) *Occurrence* — Cholesterol is widely present in the human beings, animals and also in the plant tissues. It exists as free form as well as in ester form but the proportion of these two forms is different in different organs. It remains in free form in bile. Body fluids all over the body contain cholesterol *except cerebrospinal fluid* where it is present in trace amount. Cholesterol is present in adrenal gland, glandular tissues, spinal cord, muscles, liver, kidney, cardiac muscles, blood *etc.* Blood contains high amount and may be about 17 per cent. Cholesterol in blood varies from 150 to 200 mg per 100 ml and its distribution is more or less same in the corpuscles and the plasma. Blood cholesterol level may increase to some extent during pregnancy but this is not of concern as the cholesterol level becomes normal after child-birth.

Cholesterol is present in all cells, both in the cytoplasm and cell-membrane. By its present in cell-membrane, it maintains structural integrity of cells.

(ii) *Functions* — Cholesterol acts as the starting point for synthesis of steroid hormone, vitamin D, bile acid, cholic acid, sex hormones, adrenal cortex etc. Fat is transported through blood as esters of cholesterol. It prevents red blood cells from breaking down, *i.e.,* prevents auto-haemolysis. Protection against infective organisms in the body is an important role of this sterol. Due to its presence in the cell-membrane, it maintains permeability of cells. Cholesterol plays part in cell division also (Although cholesterol is much blamed for atherosclerosis, it has also many good effects in the body).

(iii) *Chemical aspects* – Cholesterol, other name of which is cholesteryl alcohol or cholestenin has it IUPAC name as (3-beta)-cholest-5-en-3-ol, is a monohydric, secondary alcohol. It is white, crystalline and the crystals are rhombic or rectangular in shape. Cholesterol is soluble in organic chemical solvents like ether, chloroform, acetone *etc.* and not in water. Wax is formed when it is mixed with fatty acids and by that, it absorbs much water. Structure of cholesterol is presented in Figure 14.

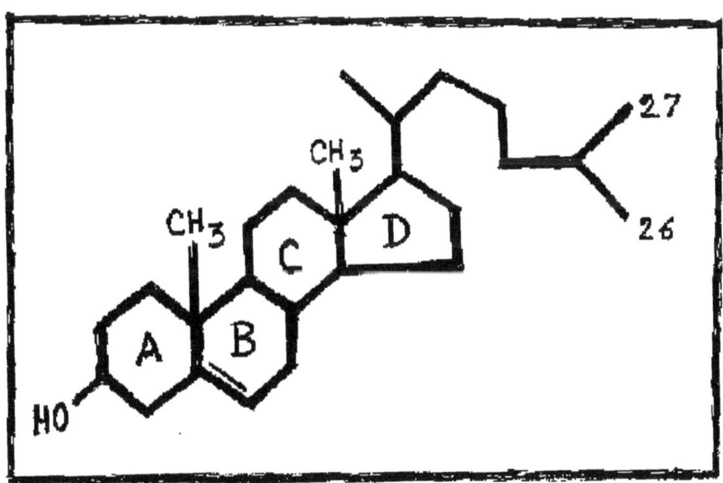

Figure 14. Cholesterol (C-27 structure).

(iv) *Synthesis* – Cholesterol synthesis is a highly complex process and occurs through many steps. *In the first step, acetate is activated* to acetyl CoA. Two molecules of acetyl CoA are then condensed together to give rise to aceto-acetyl CoA. From this, *mevalonic acid* is formed. This is phosphorylated by ATP and diphospho-mevalonic acid is formed. On undergoing series of changes, it gives rise to *lanosterol* and this gives rise to cholesterol.

(v) *Degradation* – It has been known that although the cyclo-pentano-perhydro-phenanthrene (CPP ring) is synthestized in the body of the human beings and animals but it could not be degraded to carbon dioxide and water. Removal of the CCC ring may be done by conversion of cholesterol into bile acids and then removed from the gut. Some part of cholesterol may be removed on conversion of coprostanol and cholestanol and some part may be converted in steroid hormone in the endocrine gland. There may be other mechanisms of removal of cholesterol.

(vi) *Absorption* – Fat has its necessity for absorption of cholesterol from intestine. For moving into lymphatics, cholesterol is esterified with fatty acids in the intestine.

(B) Dehydro-cholesterol

It is present in the skin and is converted to vitamin D on exposure to ultra-violet ray of the sun.

(C) Ergosterol

It is present in cell-membranes of yeast and some other fungal species, *e.g., Claviceps purpurea*. On exposure to ultra-violet ray, this is converted to vitamin D.

(D) Calcipherol

It is a synthetic vitamin D, derived from ergosterol. (Refer to the Chapter 5 on Vitamins).

(E) Bile Acids

There are two types of bile acids : primary and secondary. Primary bile acids are produced from cholesterol in liver cells and examples are cholic acid and cheno-deoxy-cholic acid. Secondary bile acids are deoxy-cholic and litho-cholic acids. These are formed in intestine when intestinal micro-organisms act on the primary bile acids. Proportion of cholic acid may be upto 60 per cent of the total and proportion of de-oxy cholic acid is 25 per cent while that of cheno-deoxy-cholic acid may be upto 20 per cent.

(F) Sex Hormones (hormone = *I arouse to activity*)

Male sex-hormones are testosterone and androsterone while female sex-hormones are estrogen and progesterone. There are stated in the following:

(i) *Testosterone* – This is an androgen of the testes of humans and animals. This is produced by the *adrenal cortex* in both males and females of human beings. It stimulates growth of secondary sexual characteristics and blood flow.

(ii) *Androsterone* – This is an androgenic steroid. (Androgen stimulates male characteristics). Androsterone is a metabolite of testosterone and androstenedione.

(iii) *Estrogen* – It induces development of female sex characteristics, particularly that produced by the ovary. Estrogen may be natural or artificial. Natural estogens are estradiole, estrone and other metabolic end products.

(iv) *Progesterone* – This is obtained from corpus luteum and placenta. It is responsible for menstrual cycle and may be used for post-menopausal syndrome. The sex hormones (Figure 15) do not have side-chains in position 17.

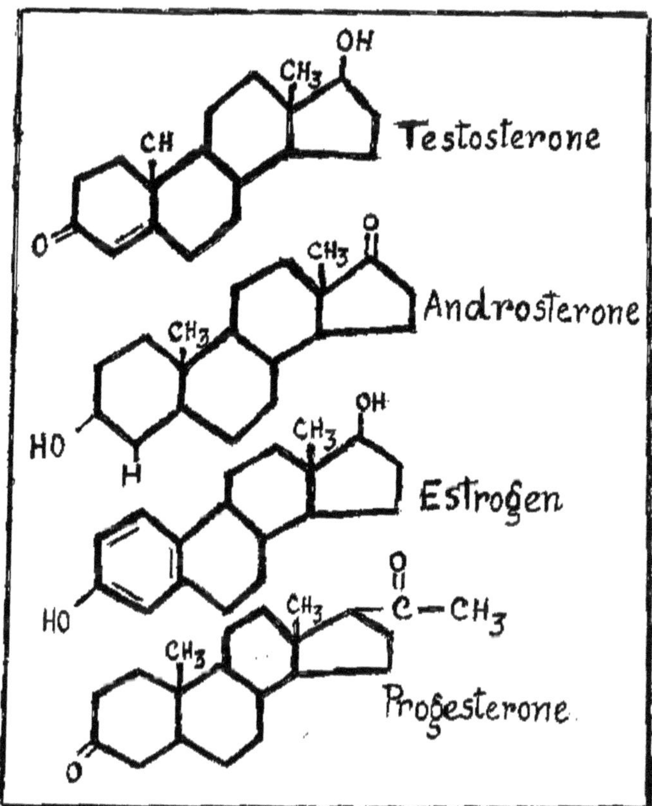

Figure 15. Male and Female Sex-Hormones.

4.8. Digestion of Lipids

Saliva in the mouth cavity contains enzymes, *viz.*, salivary amylase (= ptyalin) and also lingual lipase, which are produced in the undersurface of the tongue. Salivary enzymes cannot digest lipid but *lingual lipase* can digest. But much digestion of lipid is not done by the lingual lipase.

Lipase in gastric juice can digest lipid to some extent. Lipase in the pancreas is responsible for digestion of major part of the lipids. But it is an essential feature that the fat molecules are needed to be split off in extremely fine particles for the pancreatic enzymes to act on them. Breaking lipid molecules into very fine particles is called *emulsion*.

The lipid molecules come into emulsified form when they are acted upon by bile salts, phospholipids and fatty acids at neutral pH. On digestion, a lipid molecule releases to molecules of fatty acids with one molecule of monoglyceride. Digestion of about 98 per cent of lipids is done in the small intestine.

4.9. Adverse Effects of Lipids

Despite many beneficial effects on health rendered by fats, oils and many other lipids, at the present time, the nutritional scientists are of the opinion that firstly, too much consumption of lipids is harmful to health and secondly and more importantly, some types of fats or lipids act upon cardiovascular and nervous systems to such an extent that may kill a person or make him or her handicapped in life.

Cholesterol which is stated earlier is particularly blamed for these life-threatening maladies. Cholesterol molecules form esters by reacting with fatty acids and the esters get deposited within the layers of aorta (= main trunk of the arterial system), carotid (= vessel that comprises the principal blood supply to the head and neck), coronary arteries (= which supply blood to the heart), cerebral arteries (= which supply blood to the brain) and other arteries. Due to such ester deposition, the intima of the artery (= innermost layer of the walls of the arteries) is raised and becomes narrowed, hardened and their *elasticity is lost*. This results in reducing flow of blood through the arteries, *i.e.*, sufficient blood is not supplied to the heart or to the brain. These abnormalities are respectively referred to as cardiac or cerebral ischaemia. As a result of less flow of blood, the organs become non-functional and dead.

Nutritional scientists are of the opinion that intake of certain types of fats and oils in particular, may lead to such abnormalities (ischaemic condition) to a greater extent.

It has been known that regular intake of saturated fats (containing high amount of *saturated fatty acids*) and in high amount tend to increase blood cholesterol level. Accordingly, fats of this kind which are obtained from animal source foods, such as mutton, beef, pork, butter *etc.* and also from some plant source oil, *e.g.*, coconut oil are highly conducive to raise blood cholesterol level.

Among saturated fats, those which are regarded as polyunsaturated fats tend to lower total cholesterol level of blood. However, it is believed that they also tend to lower that part of cholesterol which is beneficial, *i.e.*, high-density lipid cholesterol (HDL - cholesterol). Oils obtained from safflower and sunflower which are much consumed in the central part of India are examples of this group.

Monounsaturated fats, *e.g.*, oils obtained from *canola* and *olive* are on the other hand, very good in that, they greatly reduce the cholesterol level but do not reduce the good cholesterol, *i.e.*, HDL-cholesterol.

High consumption of fats may also lead to other health abnormalities. Lipemia is a condition in which, abnormal amount of fat gets deposited in the blood. Lipedema is another abnormality, in which skin at the lower part may get afflicted due to high accumulation of fat. Lipidemia is also an abnormality and the syndrome is associated with painful accumulation of fat in the legs. Fat accumulation may take place around the ankles.

Obesity (= corpulence) has assumed a serious proportion all over the world and India is one of the leading countries. Even child obesity has been very serious now. Overeating particularly saturated food and animal source food and less expenditure of energy have been stated to be the cause. Obesity is not a cosmetic problem only but leads to many metabolic diseases. Severe restriction of saturated food is advocated to avert the situation. It should be remembered that Body Mass Index is not always a measure to adjudge obesity of a person.

Despite the fact that consumption of some types of lipid foods has been considered to be responsible for the above health abnormalities, further researches are necessary to confirm the concept. Consumption of fat and particularly saturated fats is not always found to bring about the abnormalities. It may be mentioned that a pig which is only fed with grasses and leaves is found to be very much fatty within a few months after birth. In fact, genetic factor in an individual impairs fat metabolism and fat addition in the body is higher when high amount of saturated fat is consumed by a person, who for genetic reason cannot breakdown fat in the body for which, the fat gets accumulated.

Lastly, it may be commented upon that a large section of physiologists and nutritionists at the present time is of the opinion that severe restriction of fat in the diet may even be harmful for functioning of heart, brain and other vital organs in the body. A minimum amount of this nutrient is necessary for good health. Further, it should be noted that not only high consumption of fat but improper metabolism of fat may bring about such disorders.

4.10. Fat Content of some Food-stuffs

Fat content of some food-stuffs as listed by Gopalan *et al.* (2018) in their book, "Nutritive Value of India Foods," (Revised by Narasinga Rao *et al.*) and published by the National Institute of Nutrition (I.C.M.R.) in Hyderabad has been presented in Table 13. The amount of fat present in some selected food-stuffs have only been presented. For the amount in other food-stuffs, the original title may be consulted.

Table 13. Fat Content of some Food-stuffs under different Groups (Figures are in gram per 100 gram of edible portion)

Cereal Grains			
Bajra	5.0	Sanwa millet	4.4
Italian millet	4.3	Maize (dry)	3.6
Jowar	1.9	Wheat flour (whole)	1.7
Ragi	1.3	Rice, flakes (Chira)	1.2
Rice (raw, handpounded)	1.0		
Pulses			
Soybean	19.5	Bengal gram (daal)	1.7
Black gram (daal)	1.2		
Leafy vegetables			
Drumstick (leaves)	1.7	Bengal gram leaves	1.4
Bak phool (petals)	1.4	Kakmachi	1.0
Roots and tubers			
Mango-ginger (Aam-ada)	0.7	Sweet potato	0.3
Parsnip	0.3		
Other vegetables			
Bitter gourd (small)	1.0	Drumstick flowers	0.8
Field bean	0.7		
Nuts and oilseeds			
Walnut	64.5	Almond	58.9
Pistachio nut	53.5	Sunflower seeds	52.1
Chilgoza	49.3	Cashewnut (kernel)	46.9
Gingely seeds	43.3	Niger seeds	39.0
Mustard seeds	39.7		
Condiments and spices			
Nutmeg (fruit)	36.4	Mace	24.4
Jowan (ajowan, omum)	21.8	Cumin seeds	15.0
Fruits			
Avocado	22.8	Wood-apple	3.7
Sapota	1.1		
Fishes and sea food			
Hilsa	19.4	Crab (small)	9.8
Engalu (Poruva)	9.6	Bhagon	8.8
Koi	8.8	Bachha	5.6
Ain (Mala)	7.8	Mahasole	2.3

Meat and poultry			
Egg (duck)	13.7	Egg (hen)	13.3
Mutton	13.3	Goat (meat)	3.6
Liver (sheep)	7.5	Beef	10.3
Milk and milk products			
Milk, buffalo's	6.5	Milk, cow's	4.1
Milk, goat's	4.5	Milk, human	3.4
Curd, cow's milk	4.0	Cheese	25.1
Chhana (Cow's milk)	20.8	Chhana (buffalo's milk)	23.0
Fats and oils			
Ghee (buffalo's milk)	100.0	Ghee (cow's milk)	100.0
Hydrogenated oil	100.0	Butter	81.0

Plant-source fats are mostly unsaturated.

Chapter 5

Vitamins

5.1. General Aspects

Vitamins (vita to be pronounced as vital) are *one of the six components of nutrients in the body*. These nutrients do not directly provide any sort of energy in the body but have indispensable role on nutrition as well as therapeutics and is no less important than the other nutritional components, *i.e.*, carbohydrate, protein, fat, minerals and water.

Absence of vitamins seriously hampers normal physiological functioning in the body and adversely effectuate growth, physical and mental activities and disease resistance capacity. Such condition in the body is referred to as *avitaminosis* and produces undesirable symptoms rapidly or at later time.

The term, vitamin was derived from the Latin word, *vita* which means life and suffixing the word, amine with it, which indicates amino acid. The name was proposed considering the fact that, vitamins were chemically amino acids. However, the term is unfortunate as it was known later that all the vitamins had not been in fact, amino acids. Nevertheless, the term, vitamin is still retained.

It may be mentioned in this connection that the earlier nutritional scientists had perhaps missed to attach as much attention on the study and research on vitamins as had been paid by them to the principal nutritive (= proximate) constituents like protein, carbohydrate and fat. There may be three possible reasons for this gap. These are, (i) vitamins do not provide energy, at least in a direct manner, (ii) they are required in small amount and some even in extremely small amount and hence, *many of the vitamins were considered insignificant* and (iii) some of the vitamins and the quantity required by them are mostly obtained from ordinary *staple diet* taken by a person. Thus, a section of nutritionists even regarded that vitamins had although necessity in the body but of secondary rather than of primary importance.

With passage of time when the science of physiology was greatly explored, importance and indispensability of the various vitamins came to the knowledge of the scientists. As a matter of fact, the following factors in particular, could be attributed which led the nutritionists, physiologists, biochemists and the physicians to pay serious attention on these components of nutrients. These are, firstly, *metabolic activity, growth* and *maintenance of health* are seriously hampered in their deficiency. Secondly, many of the vitamins have a crucial role in *utilization of protein, fat* and *carbohydrates* besides that, they also play a great role in regard to release of energy from the said three components. Thirdly, many of the vitamins are although required to the body in small or extremely small amount, the amount that becomes necessary had indispensability, for *lack of that amount gives rise to serious health hazard*. Fourthly, many of the vitamins and their required amount are though obtainable from the common food, it does not happen so always. The reason is, foods that are taken by people always do not provide all the vitamins or the amount of each vitamin which actually becomes necessary. It may also be brought to notice that apart from nutritional importance, almost all the vitamins have now been known to have *therapeutic value*, not only to prevent but also to cure a number of ailments.

Thus, the science of vitamins has assumed a great role in physiology and medicine. In accordance with, studies and researches on vitamins are being carried on now extensively in different parts of the world.

5.1.1. Definition

Vitamins comprise as one of the six components of nutrients in the body which are indispensable for growth, metabolism and thus, maintenance of health. These are complex and potent chemical compounds or related set of molecules which are obtained *only from organic substances*. Unlike other nutritional components like protein, carbohydrate and fat, vitamins are required in the body in small and some in extremely small amount.

Requirement of some vitamins is *conditional*. This is in view of the fact that it depends on *particular organism(s)* and on *circumstances(s)*. Thus, a vitamin, *e.g.*, vitamin C may be cited which must be provided to the body of human beings, apes, bats and some other animal species from external sources (food), because they cannot synthesize this vitamin in their body. On contrary, animals like tiger, lion, elephant and many other animals are able to synthesize vitamin C in their own body and hence, it is not necessary for these animals to get this vitamin from external source. Some other vitamins like biotin and vitamin D may be mentioned which are required to the human beings under special circumstances.

Vitamers

A particular vitamin may refer to more than one compound in its group. All these chemical compounds show *similar biological activity* as with the respective vitamin. These compounds which are chemically somewhat different but are similar in activity of the concerned vitamin are termed as *vitamers*. Vitamers belonging to the particular vitamin are *not chemically same with that vitamin* although may have *identical or related chemical structures*. However, their *biological activity is same with that (generic) vitamin*. The vitamers may be converted to the active form of the vitamin.

A list of generic descriptor names of vitamins and vitamer chemical names or chemical class of compounds are stated below:

Vitamin A: Retinol, retinal, carotene *etc.*

Vitamin B1 : Thiamine, thiamine monophosphate *etc.*

Vitamin B2 : Riboflavin, flavin adenine dinucleotide *etc.*

Vitamin B3 : Niacin, niacinamide *etc.*

Vitamin B5 : Pantothenic acid, panthenol *etc.*

Vitamin B6 : Pyridoxine, pyridoxamine *etc.*

Vitamin B9 : Folic acid, folinic acid *etc.*

Vitamin B12 : Cyanocobalamin, methyl cobalamin *etc.*

Vitamin C : Ascorbic acid, calcium ascorbate, dehydro-ascorbic acid *etc.*

Vitamin D : Calcitriol, ergocalciferol (D2), cholecalciferol (D3) *etc.*

Vitamin E : Tocopherols, tocotrienols *etc.*

Vitamin K : Phylloquinone (K1), menaquinones (K2) *etc.*

Antivitamins

Antivitamins are chemical substances that *antagonize the action of vitamins*. That is, these are chemical compounds which counteract the utilization of vitamins. If any antivitamin is present in the body, it will suppress the action of that vitamin to which it counteracts. Accordingly, deficiency symptoms of that vitamin will be produced even if that vitamin is adequately present in the body or is administered through oral route or instilled in the blood stream. A specific antivitamin compound will counteract the activity of a particular vitamin or its vitamer(s). Some examples of antivitamins are given below:

Deoxypyridoxine is a compound which antagonizes the vitamin, B-6. Similarly, 3-acetyl pyridine and pyridine sulphonic acid counteract the activty of another vitamin of the B group, which is niacin. Vitamin B-2 is antagonized by galactoflavin and another B vitamin, folate is antagonized by the compound, aminopterin. There are many other compounds, each of which antagonizes the effect of some specific vitamin(s). It is believed that even tobacco smoking or any other habit of tobacco antagonizes the effect of vitamin C.

5.1.2. Discovery

The vitamins belonging to various types were discovered after series of studies undertaken by the chemists and biologists. These studies were conducted by the process of feeding and elimination of different foods to the laboratory animals. Sometimes human subjects were also used and mention may be made to the Scottish surgeon of the British Navy, Captain James Lind, who discovered that feeding oranges and other citrus fruits to the sailors had remarkably prevented the scurvy disease in them when they sailed in the ships for months together.

A number of vitamins had been isolated and chemically identified at the later part of the 18th century. Rickets in rats had been prevented by feeding fish oil and the active substance of the fish oil had been termed as *anti-rachitic vitamin A*. (However, it was learnt later that the active substance was actually vitamin D and not A). The scurvy disease in mice was studied by the Soviet surgeon, Nikolai Lunin by feeding them with an artificial mixture made with milk constituents. It may be mentioned that curing of night blindness in human beings by feeding them liver preparation had been a practice among the ancient Egyptians.

The beriberi disease drew attention of the scientists which was common among people in the eastern Asia who had the habit of consuming much amount of white, *polished rice*. The physician, Takaki Kanehiro in 1884 studied that some deficiency in their food was responsible for this disease. Later, Christian Eijkman in 1897 discovered that in chickens, the disease could be prevented by feeding unpolished rice. Frederick Hopkins later put forward that some *accessory factors* had been present in many foods and these were greatly responsible to prevent many diseases.

The Japanese scientist, Umetaro Suzuki first isolated *vitamin B complex* in 1910 from an aqueous extract made from *rice bran* and he termed it *aberic acid*. Within a year, the Polish biochemist, Kazimieriz Funk of the Lister Institute of London in 1911 isolated the active substance, *i.e.*, the anti-beriberi factor in pure form. He opined that the factor was chemically, an *amine*. Hence, as suggested by Max Nierensteine, the term "vitamine" was proposed by him. Such a term was proposed on consideration that the *amine which was conjectured had been of vital necessity to health* and hence, *should be termed as vitamine, i.e., a contraction of two words, vital and amine.*

Lunin and Bunge in 1888 put forward that apart from protein, carbohydrate and fat, some other factors were also there which had great role to health and these should be obtained from food. However, the term, vitamine which was proposed by Funk and Nirenstein was found to have some incorrectness. That is, the active substances which remained present in food and were essential to maintain health *were all not actually amines* in their chemical form, although some were amines. Accordingly, J. C. Drummond in 1920 suggested that although the active substances were vital (= essential) to health and all of them were not chemically amines, the term, *vitamine should be termed as "vitamin" and the last letter, "e" should be dropped.* Accordingly, the term, vitamin is now universally used.

Discovery of the different vitamins in respect of years and the scientists have been stated in the following:

- The term vitamine was proposed by Kazimierz Funk and Max Nierensteine in 1911.
- Kazimierz Funk discovered *thiamine* in 1912.
- T. Smith and E. G. Hendrick discovered *riboflavin* in 1926.
- Conrad Elvehjem in 1937 discovered *nicotinic acid.*
- Lucy Wills in 1933 discovered *folic acid.*
- Paul Gyorgy in 1934 discovered *vitamin B-6.*
- James Lind in 1747 discovered *vitamin C.*
- Edward Mellanby in 1922 discovered vitamin D.
- Herbert Evans and Katherine Bishop discovered *vitamin E* in 1922.
- Henrick Dam discovered *vitamin K* in 1920.

5.1.3. Classification

Vitamins which have been known so far are of diverse types. They are usually *named according to English alphabets, e.g.,* vitamin A, B, C, D and in this way. But it should be noted that all the vitamins are not named according to alphabets. Again, alphabetical nomenclature differs according to authors as the same vitamin is sometimes designated with different alphabets. Another point should be noted that, in alphabetic nomenclature, there are some *gaps* and this is particularly common in the B - group of vitamins. The reason for such a gap is due to the fact that a vitamin which was assigned an alphabetical name was later found to be not a true vitamin and hence, that vitamin was *dropped* from the list, *i.e.,* from the series of vitamins. Further, some authors have given names of vitamins just arbitrarily. For example, vitamin K was named from the German word, *koagulation*, which in English is spelt as coagulation. Again, the vitamin P was named from the paprika as it was first isolated from fruits of paprika plant.

For human *beings, in general, thirteen types of vitamins are known to be required.* But it could be mentioned that the actual number that is required is known to vary according to authors. Also that, many of the vitamins have been stated to be required *conditionally*, which means that they are required according to specific condition, *i.e.,* circumstances.

However, the different vitamins have been *broadly classified under two groups in accordance with whether they are soluble in fatty substances or in water.* There are stated as follows:

(i) Fat-soluble Vitamins

These vitamins are soluble in fat, oil or other lipid (fatty) substances, or, in such organic chemical substances, *e.g.*, ether, acetone *etc.* known as fat-solvents, in which fats and other lipid substances are soluble. Four types of vitamins come under this group. These are, vitamins A, D, E and K.

(ii) Water-soluble Vitamins

Vitamins belonging to this group are soluble in water. Vitamins of the B-group and the vitamin C come to this group. It is mentioned in this context that vitamin B was later known to be of *several types* and not only one compound. All these compounds behave as vitamin B in their activity and all of them are soluble in water. Hence, all these vitamins are considered as B vitamins and there are *collectively designated as Vitamin B Complex*. The different compounds of the B-group of vitamins, *i.e.*, B-complex vitamins are stated below:

Thiamine (= vitamin B1), Riboflavin (= B2), Nicotinic acid (= Niacin, B3), Pantothenic acid (= B5), Pyridoxine (= B6), Folic acid (= B9), Cyanocobalamin (=B-12), Lipoic acid, Choline, Lipotropic factor, Biotin (=B7, Vitamin H or Co-enzyme-R), Para-amino benzoic acid, Inositol, Vitamin B4 *etc.*

It is mentioned in this connection that apart from the B group of vitamins and the vitamin C, antother vitamin, named as vitamin P was also taken into the list of the water-soluble vitamins. It was named vitamin P as it was first isolated from paprika (P from paprika). At present, most of the authorities on vitamins are however, of the opinion that vitamin P should not be considered as a vitamin. The committee on nomenclature of vitamins consisted of the American Society of Biological Chemists and the American Institute of Nutrition recommended (1950) that the name vitamin P be dropped. Instead of that, the term, *bioflavonoids* had been proposed to the group of compounds, which should be apposite.

5.1.4. Physiological Functions

Vitamins are not constituents of any chemical compound in the body but *they play essential role to serve as co-enzymes of many enzymes in the body.* Many have role in prevention and cure of some diseases as well.

They are obtained in the body from food or other external source but some are synthesized in the body itself. For example, vitamin A is synthesized from the pigment, carotene (= provitamin A) and vitamin D is synthesized from ergosterol on exposure to ultra-violet ray. Many micro-organisms that inhabit in the gastro-intestinal tract are also able to synthesize some vitamins of the B group.

Vitamin C and E have powerful *anti-oxidant properties* withoud producing side-effects. Carotene which is the precursor of vitamin A (provitamin A) has also anti-oxidant properties. Vitamin D has *hormonal role* in the regulation of mineral metabolism for bones and some other organs. C. C. Chatterjee (2004) in his book, "Human Physiology," Vol. 1 (Medical Allied Agency, Kolkata - 700 009) has made a presentable summary of the vitamins that act on different systems, which is stated below:

(1) *Nervous system* – (i) Vitamin A. (ii) Vitamin B- thiamine, nicotinic acid, pantothenic acid, pyridoxine, choline. (iii) Vitamin D - acts indirectly through its action on bones.

(2) *Alimentary system* – (i) Vitamin A – on the epithelium, glands and secretions. (ii) Vitamin B - thiamine (tone, appetite and secretion), riboflavin (mouth ulcer), nicotinic acid (gastro-intestinal disorders in pellagra). (iii) Vitamin C - malformation of teeth, haemorrage from gums, intestine and increased susceptibility to infection of the gastro-intestinal tract, *etc.*

(3) *Circulatory system and blood* – (i) Vitamin B - thiamine (cardiac damage of beriberi, incerased blood lactate), folic acid, cyanocobalamin, pyridoxine. (ii) Vitamin C. (iii) Vitamin D (on blood calcium). (iv) Vitamin K. (v) Vitamin P.

(4) *Bone formation* – (i) Vitamin A. (ii) Vitamin C. (iii) Vitamin D.

(5) Epithelium, skin and hair – (i) Vitamin A. (ii) Vitamin B – riboflavin, nicotinic acid, biotin, pyridoxine, inositol, para-amino benzoic acid. (iii) Vitamin C.

(6) *Reproductive system* – (i) Vitamin A. (ii) Vitamin B - pyridoxine, folic acid, pantothenic acid, vitamin B-12. (iii) Vitamin C. (iv) Vitamin E.

(7) *Growth* – (i) Vitamin A. (ii) Vitamin B - Complex. (iii) Vitamin C. (iv) Vitamin D.

5.1.5. Absorption

All vitamins are absorbed in the gastro-intestinal tract except vitamin D which is formed under the skin when exposed to ultra-violet ray (morning sun). Vitamin B-12 is absorbed in the ileum and the rest are mostly absorbed in the jejunum.

5.1.6. Storage

The fat-soluble vitamins, *i.e.*, vitamin A, D, E and K are stored in the body in *liver* for long period. They are released by the liver gradually and regularly when there is requirement in the body. The water-soluble vitamins are not stored and the excess amount is expelled out mainly through urine and also faeces and sweat. However, interestingly, vitamin B-12 which is a water-soluble vitamin is *stored in the liver*. This vitamin may be stored in the liver for even three years or more, if necessity is not felt. Another water-soluble vitamin, *i.e.*, vitamin C is also stored in the adrenal cortex for a considerable period under certain conditions.

The specific influences of the different fat-soluble and water-soluble vitamins on human health have been individually discussed under the following heads.

5.2. Vitamin A

Vitamin A is considered as the most important of the four fat-soluble vitamins. It comprises a group of unsaturated organic compounds of nutritional importance which includes the compounds, *viz., retinol, retinal, retinoic acid and also the plant pigment, carotene*. Vitamin A exists in *cis-* and *trans-* forms and the latter form occurs naturally. Vitamin A was discovered by Mc Collum and Davis in 1915 from constituents obtained from egg and milk products.

5.2.1. Functions

A number of physiological functions of vitamin A are known and the pertinent ones are as follows:

(i) Most important role of this vitamin is *protection of eyes* from serious diseases.

(ii) For protecting the body against many pathogenic infections, vitamin A is given the status as *anti-infective vitamin*. Degeneration of epithelial tissues which makes them liable to be easily infected is slowed down by this vitamin.

(iii) For proper functioning and structure of bones, the vitamin A has great role.

(iv) By retarding degeneration of nerve cells, it maintains healthy nervous system.

(v) Activity of epithelial tissues (= tissues at the outer surface of the body and the lining of body cavities) is maintained by this vitamin.

(vi) It plays part in the formation of muco-polysaccharides.

(vii) In protein synthesis, it plays part.

(viii) In stabilising permeability of mitochondria, it helps in respiratory activity.

(ix) Stability of cell-membrane is maintained by this vitamin.

(x) It has role in preventing urinary stone formation.

(xi) Ovulation in rats is observed to be improved by vitamin A although this is not confirmed in case of human beings.

5.2.2. Principal Food Sources

Vitamin A is obtained by the body from both animal as well as plant source foods. When it is obtained from animal source food, it is an *already formed vitamin A, i.e.*, chemically it is retinol, or its aldehyde form, retinal. This is abundantly present in the oil of the liver of sea fishes and is richest in the liver oil of halibut fish. Retinol or retinal is also rich in butter, cheese and egg-yolk.

When it is obtained from plant source food, it is not actually vitamin A but an yellow or orange-coloured pigment which is known as *carotene*. Vitamin A, *i.e.*

retinol is formed from this pigment in the body and hence, it is called *provitamin A*. Carrot is a rich source of carotene and hence, the name, *carotene is* derived from that but among all foods, carotene is richest in ripe pulp of mango fruit when the pulp turns into orange or yellowish orange in colour.

Carotene exists in *three isomeric forms*, which are, alpha-, beta- and gamma-forms of carotene. Another related compound, xanthophyll pigment, *i.e.*, beta-cryptoxanthin may be named which is abundantly present in yellow-orange or orange-red colour pulp of ripe papaya fruit. All these pigments that are obtainable from plants give rise to vitamin A when they go inside the body. From the chemical point of view, one molecule of the *beta* form of carotene gives rise to two molecules of vitamin A, *i.e.*, retinol while one molecule of each of the other forms of the carotene pigment gives rise to one molecule of vitamin A. The chemical explanation of this occurrence has been stated under the next heading.

5.2.3. Chemical Aspects

Vitamin A, which is an unsaturated form of primary alcohol has a typical ring structure, which is known as *beta-ionone ring* and this is attached to a side chain of hydrocarbon that terminates in an alcohol group. *This is called retinol and the aldehydic form is called retinal.*

The carotene pigments (= provitamin A) have one or more beta-ionone rings and these are joined by long hydrocarbon chains. In beta-carotene, there are two rings of beta-ionone while the alpha- and the gamma-carotenes have one beta-ionone ring in each. For this reason, *one molecule of beta-carotene produces two molecules of vitamin A (= retinol) while one molecule of alpha-carotene or gamma-carotene gives rise to only one molecule of vitamin A*. However, although theoretically this is true, it does not happen practically due to physiological inefficiency and in fact, one molecule of beta-carotene gives rise to one molecule of vitamin A only when goes inside the body. Vitamin A which is derived from the pigment, carotene (= provitamin A) exists in two forms, *viz.*, vitamin A1 and vitamin A2. These are closely related to each other.

Vitamin A is apparently a colourless, oily substance but on suitable fractionation, it gives pale yellow coloured needles. Under anaerobic condition, it is heat-stable. Its activity in food is partly reduced when it is fried. Exposure to ultra-violet ray destroys its activity quickly.

5.2.4. Absorption

Retinol gets mixed with the bile salt in the intestine. It forms esters on reaction with fatty acids in the mucous membrance of the intestine and then transported to the liver. Beta-carotene, which is the pro-vitamin A converts into vitamin A, *i.e.*, retinol in the intestinal mucous membrane, although some part of the beta-carotene is absorbed without undergoing conversion into vitamin A.

5.2.5. Storage

Major part of the vitamin A, *i.e.*, 90 to 95 per cent is stored in the liver in the form of esters of fatty acids. Remaining part is stored in kidneys, lungs and very few proportion in other organs.

5.2.6. Deficiency Symptoms

Vitamin A is stored in the liver and this stored amount is released when there is necessity of it in the body. Hence, deficiency symptoms of vitamin A occurs when the amount stored in the liver goes very low but the necessity in the body is high. This takes place when there is a chronic deficit of this vitamin in the daily diet of a person or its absorption in the body is hampered. However, deficiency of vitamin A is common in people of different countries and also in many parts of India. Some of the common deficiency symptoms may be stated as follows:

(i) Night blindness, which is known as *nyctalpia* is a direct effect of vitamin A deficiency and earliest symptom. However, hereditary factors and too much smoking are also known to be possible causes for this abnormality. Early symptoms of this disorder is inability to see in dim light and this usually happens after exposure to bright light.

Night blindness occurs in a person due to loss of *rhodopsin* in the red cells of the retina in the eye. Rhodopsin (formerly known as visual purple) is a chromo-protein (= coloured protein) which contains the pigment, retinal as the prosthetic group of the protein, which is *opsin*. On exposure to light, the protein breaks up and the retinal part is freed. In presence of light, rhodopsin is transformed to lumirhod-opsin and this is transformed to metarhod-opsin and then, a yellow-coloured combined form, *i.e., trans-* form of retinal and opsin are formed. In the blue part of the light spectrum in the laboratory glass-plate, the *trans-* form is changed to *cis-* form. The *trans-*retinal is changed to *trans-* form of vitamin A in the eye by the enzyme, reductase. When rhodopsin is synthesized in the dim light, the *cis-* form of vitamin A which is active enters retina through blood and this is oxidized by enzyme. The *cis-* form gives rise to rhodopsin with combination of opsin.

Wald put forward the mechanism of involvement of vitamin A to normal vision as a *cyclic phenomenon*. This occurs due to continuous degradation (= breakdown) and synthesis of rhodopsin in the retina of the eye. Hence, the cyclic process is recognized as *Wald's Vision Cycle*.

(ii) *Xerophthalmia* (syn. xeroma, xeropthalamus) is another disorder of the eye which occurs due to vitamin A deficiency. In this case, corneal epithelium of the eyes becomes abnormally dry (xerox = dry), thickened and the eyes become red, encrusted and haemorrhagic. The eye-ball becomes lustreless and the cornea is degenerated. In severe cases, the cornea is ulcerated and develops the condition known as *keratomalacia*, which means softening of the eyes. If untreated, the disorder may result in complete blindness.

(iii) Vitamin A deficiency results in *phrenoderma* in skin, which goes by the term, *toad skin*. The skin becomes thickened, dry keratinized and develops popular eruption.

(iv) Urinary stone formation may result due to vitamin A deficiency as the epithelium gets degenerated.

(v) Degeneration of the respiratory tract epithelium may also occur due to vitamin A deficiency.

(vi) Vitamin A deficiency may result in degeneration of the glands in the gut (= gastro-intestinal organs).

(vii) Abnormal growth of bones particularly in the vertebral column and skull may take place in some persons due to deficiency of this vitamin.

(viii) Degeneration of the nervous tissues may be brought about due to deficiency.

(ix) Deficiency of vitamin A may cause diarrhoea in some people.

5.2.7. Daily Requirement

Recommended dietary allowance of vitamin A to the people belonging to various groups has been stated in Table 17 under the Chapter 9 (Dietetics). The actual requirement of vitamin A differs on many factors, according to persons and conditions.

5.2.8. Adverse Effects

Despite the fact that vitamin A, the eye-protective or anti-infective vitamin has a number of health-giving benefits, excess accumulation of it in the body renders great harm. This happens when a person consumes very high amount of vitamin A or, carotene which is precursor of vitamin A regularly and for a long period. High rate of absorption of vitamin A in the body system may also result high accumulation in the body, *i.e.*, in the liver where this vitamin is adequately stored. Excess accumulation of vitamin A results in unhealthful condition which is clinically known as *vitamin A-hypervitaminosis*. Such a condition brings about syndromes like heavy loss of weight, loss of plentiful hairs, atrophy (= death and resorption of cells, lowered cellular proliferation) of skin, ulcer in the eye, osteoporosis, reduction of plasma prothrombin, muscular loss of vitamin C *etc.*

It is also found that if there is excess accumulation of carotene (= pro-vitamin A) in the body, an abnormality like yellow or orange patches appear on palm and other parts of the skin. The abnormality is called *carotenemia* or *xanthosis*. Sometimes it is wrongly diagnosed as jaundice. But in carotenemia, retina of eye does not look yellow and bilirubin in blood is also not high in this case like jaundice. However, the yellow patches or carotenemia (= xanthosis) although looks like patches of jaundice but it is neither jaundice nor any serious type of disorder. The patches disappear when the carotene intake in the body is restricted. People of some parts in north India are sometimes found to develop such patches when they take high amount of ripe mango fruits regularly during the season. Ripe pulp of mango is the richest source of beta-carotene among all food-stuffs.

5.3. Vitamin D

Vitamin D is another fat-soluble vitamin which is commonly known as *anti-rachitic vitamin*. It is naturally synthesized in the body of the human beings and other mammals adequately on exposure to sunlight (ultra-violet part of sunlight) and thus, it is not essential to be obtained from diet. Hence, this secosteroid compound is considered as a hormone.

Vitamin D is in fact, a group of compounds and among these, the two forms, *i.e.*, vitamin D-3 and vitamin D-2 are of importance to the human beings. Mc Collum in 1910 identified vitamin D from liver oil of sea fishes. Thereafter, existence of dehydrocholesterol in the skin of human beings and ergocalciferol in plants were discovered.

5.3.1. Functions

Most important function of vitamin D is its high rate of absorption from the intestine and *efficient utilization of calcium and phosphorus* in the body. This vitamin has indispensable role in the formation of bones and teeth. By calcium absorption, increase of citric acid content in blood and bone is accomplished by this vitamin. Vitamin D activates alkaline phosphatase in kidney, intestine and other parts in the body.

5.3.2. Principal Food Sources

The D-3 form of vitamin D is synthesized in the body when the compound, 7-dehydrocholesterol, which is present on the skin gets exposed to ultra-violet part of sunlight which is abundant in the morning sunlight. About 90 per cent of the requirement of vitamin D of human beings is obtained by the body in this way.

The D-2 form of vitamin D is abundant in the liver oil of fishes and particularly, sea fishes. It is richest in liver oil of halibut fish (about 1000 International Units per gram) and is also very high in cod liver oil. High amount is also present in egg-yolk, butter and cheese. Much amount of vitamin D is found to be present in many fungal species as well. Gabriel Stangl in 2018 from the Martin Luther University, Helle-Wittenberg in Germany reported that cocoa butter and dark chocolate had been rich source of vitamin D-2.

5.3.3. Chemical Aspects

Vitamin D is a group of compounds of which, the two principal forms are vitamin D-2 and vitamin D-3. Vitamin D-2 is also known as calciferol. This is produced when the plant sterol, *viz.*, ergosterol is exposed to ultra-violet ray which is abundant in the morning sun. This ray (= short wavelength ray) brings about *rearrangement of atoms* in the ergosterol molecule and gives rise to calciferol. The vitamin D-3 (the activated form of vitamin D-3) is produced when the sterol produced by animals, *viz.,* 7- dehydrocholesterol is exposed to ultra-violet ray. This compound remains present on the skin of human beings and other mammals.

Other forms of vitamin D are vitamin D-1, D-4, D-5 *etc.* Vitamin D-1 is a mixture of molecular compounds of ergocalciferol and lumisterol in equal proportion. Vitamin D-4 is 22- dihydroergocalciferol and vitamin D-5 is sitocalciferol, made from 7- dehydrositosterol. All forms of vitamin D are stable if not heated above 100°C.

5.3.4. Absorption

Vitamin D is absorbed in the intestine in presence of fatty acids and bile.

5.3.5. Storage

Like vitamin A, major part of vitamin D is stored in the liver.

5.3.6. Deficiency Symptoms

Deficiency of vitamin D causes great harm to the bones and teeth in particular. For the purpose of formation and maintenance of the density and formation of bones and teeth, *calcium and phosphorus ions are indispensable.* But when there is deficiency of vitamin D, *these ions are not absorbed from the intestine* and the loss of them that takes place through faeces are not compensated. As a result, deficiency of vitamin D occurs in the body and the bones and teeth are suffered.

Long-term deficiency of vitamin D results in deformity of bones, which are known as *rickets, osteomalacia* and *osteoporosis.* Rickets and osteomalacia are in fact, same disease. Rickets occur in children usually from 6 to 18 months of age and osteomalacia occurs in adults or adolescents.

In rickets (= wrickken = twisting of body parts), long bones of hand and leg *bend by the weight of the body* and the legs look like bows. Epiphyses and cartilages are not adequately calcified (= deposited with calcium) and epiphyseal line becomes wide. Bony deformation occurs in pelvis, spine and chest as well. Delayed dentition, badly formed teeth, large size of head, enlargement of liver and spleen, prominent sternum bone *etc.* are some of the other symptoms.

Osteomalacia (= osteone = bone, *malacia* = abnormal softening of tissues) is an abnormality in which bones are softened and thereby, become flexible and brittle. It is more commonly observed in pregnant and lactating women and deformity occurs in their pelvic bones. Due to the growing foetus in the womb, requirement of calcium and phosphorus is high during pregnancy as the calcium and phosphorus ions move away from the mother to the foetus, effectuating mothers to be deprived. In lactation, high calcium and phosphorus go to the new-born baby through breast milk and for this, the mother is deprived.

In osteoporosis (= porous, full of pores), calcium and phosphorus ions move away from bones to other tissues or organs where requirement is genuinely felt. By that, *i.e.*, on losing the minerals, *i.e.*, calcium and phosphorus regularly, density of the bones is greatly reduced. As a result, the bones become porous, *i.e.*, assume spongy appearance particularly at their ends. Reduction of bone density makes the bones brittle and sometimes brittleness may be so severe that even on making hand-shake with a person, the wrist bones of a person are liable to be fractured.

A type of osteoporosis is found to occur largely in women between 50 and 70 years of their age. This goes by the term as "Involuntary type of osteoporosis." Another type, known as "Senile type of osteoporosis" occurs both in men as well as women when they become more than 70 years of age.

Deficiency of vitamin D is an important cause of chronic form of respiratory illness. This is attributed for the fact that this vitamin is involved to produce the anti-microbial proteins which kill the respiratory pathogens.

Vitamin D is although largely concerned with bones and teeth, researches done in the 20th century have provided adequate evidence to convince that this fat-soluble vitamin has *many other benefits in the body*. Many gerontologists of the present day advise regulated intake of vitamin D at long interval (so that it is released slowly and regularly) to the people who are above 70 years of age. This is advised to avert ill health of bone and also other senile abnormalities in them. Of late, vitamin D is much recommended in physiological disorders of various types and many pathogenic infections. Effect of vitamin D to offer resistance against HIV virus infection is also claimed but needs confirmation.

5.3.7. Daily Requirement

Requirement of vitamin D for infants upto one year of age may be 500 – 800 I.U. (= International Units) daily and for children and adolescents, it may be 500 I.U. Requirement is felt higher in pregnancy and lactation, during which time it may be 900 I.U. or even more, particularly for those women who do not get adequate exposure to sunlight for which, vitamin D synthesis in their body is less. Many specialists in geriatrics now advise regulated intake of vitamin D to the elderly people for protection against many old-age illnesses.

It should be noted that if it becomes necessary for an individual of any age to take vitamin D as a food supplement, it should be taken at a suitable interval which may be once a week or a month depending on the potency of the supplement drug and not on every day. The vitamin D, which is taken at weekly or monthly interval as a food supplement is stored in the body and it releases the vitamin D from the stored amount gradually as required by the body. For the neonates and children, physicians may recommend everyday intake of vitamin A and D through specially formulated drugs or drops.

5.3.8. Adverse Effects

Excess intake of vitamin D in the body is harmful like excess intake of vitamin A. This is referred to as *hypervitaminosis* – D. Thus, high intake of vitamin amounting to 100, 000 I. U. daily is liable to cause hypercalcaemia, *i.e.*, excess accumulation of calcium in the blood. Symptoms of this abnormality are loss of body weight, polydipsia, polyuria, weakness, vomiting, headache, depression *etc.* Calcium deposition in the kidney and atherosclerosis may also result due to intake of very high amount of vitamin D.

5.4. Vitamin E

Vitamin E which is another fat-soluble vitamin is given much importance nowadays for its *value as an anti-oxidant rather than its role as a vitamin.* The compound is designated as *tocopherol* and its deficiencies could although be produced in the experimental animals, abnormal symptoms due to its deficiency is hardly observed in the human beings, or perhaps the role of this vitamin has not much been studied in the humans.

Vitamin E is usually referred to as *fertility vitamin* for its effect on reproduction of rats. But such an effect of vitamin E in the human beings is yet to be established. Herbert McLean and K. S. Bishop discovered the vitamin E in 1922 and it was first isolated by Evans and Emerson in 1935. Paul Karner synthesized it artificially. The name, tocopherol is derived from the Greek word which mean *child birth.*

5.4.1. Functions

Greatest role of vitamin E is its *anti-oxidant properties, i.e., removal of the free-radicals from the body,* which are accumulated in the body always and to a greater extent under certain conditions. Removal of free-radicals (Reactive Oxygen Species) greatly protect the body in many ways and it also delays the ageing process. The vitamin E is believed to have great role in muscular dystrophy, fertility, genital parts, in treating coronary diseases, hyperlipidaemia *etc.* and is also reported to have role in preventing some types of cancer. It is stated that vitamin E has some effect in preventing habitual miscarriage in women. Nowadays, vitamin E is much used for its anti-ageing effect, lustre of skin and other cosmetic purposes. Light massaging on fact with a mixture of vitamin E, calamine, groundnut oil and sandal paste is done by many to remove black spots.

5.4.2. Principal Food Sources

Vitamin E is present in considerable amount in wheat and rice germ oil, alfalfa grass, cotton seed oil, lettuce, spinach, leafy vegetables, soybean, unhusked wheat (= whole wheat), maize *etc.* Among animal source foods, egg, fish liver and muscle of animals, milk and milk products are also moderate sources.

5.4.3. Chemical Aspects

Vitamin E has several groups and each group has a *tocol nucleus.* These groups are known as alpha-, beta-, gamma- and delta-tocopherols and besides these, some other groups have also been reported. However, among all these groups, alpha-tocopherol ($C_{29}H_{50}O_2$) is most active. All tocopherol groups (forms) are highly stable in heat in absence of air (oxygen) but the activity is lost on exposure to ultra-violet ray.

5.4.4. Absorption

Tocopherols in their free form as well as in chemically combined forms are easily absorbed from the small intestine. Bile acids are however, necessary for their absorption.

5.4.5. Storage

Being a fat-soluble vitamin, major part of vitamin E is stored in the liver and some parts in other organs.

5.4.6. Deficiency Symptoms

Any definite symptom of deficiency of vitamin E is hardly noticeable in the human beings although deficiency symptoms could be produced in animals if complete deprivation of vitamin E is done through their food. In the experimental female rats, deficiency of vitamin E results in intrauterine death of foetus and in male rats, gereminal epithelium is destroyed for which, their reproductive potency is lost. Deficiency has been stated to bring about muscular dystrophy, degeneration in skeletal muscles and paralytic condition in rats. Hair growth and skin health are stated to be improved by topical application of vitamin E but this may be due to anti-oxidative property of it.

5.4.7. Daily Requirement

Daily requirement of vitamin E has not been precisely assessed but intake of 10–20 mg of the vitamin per day through diet is considered satisfactory. According to the National Institute of Nutrition (I.C.M.R.) in Hyderabad, Indian people have blood level of 0.5 mg of vitamin E per kg per ml and this is considered satisfactory, for which extra intake is not necessary. For therapeutic purpose, hair growth stimulation and particularly for use as anti-oxidant, vitamin E supplement is however, recommended by health providers.

5.4.8. Adverse Effects

Like other fat-soluble vitamins, excess accumulation of vitamin E in the body is harmful. Among other symptoms, it is believed to cause dermatitis and allergic reaction.

5.5. Vitamin K

Vitamin K, the other fat-soluble vitamin is usually referred to as *anti-haemorrhage factor* for its great role in *coagulation of blood*. For the presence of prothrombin, vitamin K coagulates blood and thereby, prevents bleeding. The name of this blood coagulation factor is derived from the German word, *koagulation*, which in English is spelt as coagulation. Vitamin K is a group of compounds. It exists in naturally occurring substances, such as phytonadione (vitamin K-1) which is present in many plants, menaquinone (vitamin K-2) which is produced by some bacteria and menadione (vitamin K-3) which is synthesized by leafy plants and intestinal bacteria. The discovery of vitamin K was done by Dam while studying haemorrhagic disorders of poultry birds.

5.5.1. Functions

Coagulation of blood is most important function of vitamin K. Uncontrolled bleeding from wound of the body takes place in its absence. The vitamin promotes synthesis of procoagulant, *i.e.*, precursor of coagulant and the factors VII, IX and X. Primary role of vitamin K is to *shorten the prothrombin time*. Vitamin K has also role in phosphorylation in the photosynthetic process in plants and phosphorylation in animal tissues. When there is loss of its activity in the mitochondria, oxidative phosphorylation is hampered. Vitamin K has important role in bones and its absence leads to osteoporosis. The deficiency also results in atherosclerosis.

5.5.2. Principal Food Sources

Vitamin K is largely obtained from plants. Cabbage, kale, alfalfa, green leafy vegetables, tomato, beet, bean, oat, nuts, many pulses, pear and some other fruits are rich sources of vitamin K. It is also synthesized by many bacteria and appreciable amount is produced by the intestinal bacteria. Yeast is a good source of this vitamin. Animal source foods are usually not good sources but milk, egg-yolk and putrified fish contain some amount of vitamin K.

5.5.3. Chemical Aspects

Vitamin K-1 is chemically identified as 2-methyl-3-phytyl-1, 4 - naphthoquinone and vitamin K-2 (obtained from bacteria) is identified as 2- methyl-3-difarnesyl-1, 4-napthoquinose.

Vitamin K-1 is a yellow coloured, crystalline oil while vitamin K2 is a yellow-coloured, solid substance. They are heat-stable but their activity is lost on contact with alcohols, alkalis, mineral acids and on exposure to light.

5.5.4. Absorption

Vitamin K is absorbed from the small intestine and bile salts are necessary for absorption like that of vitamin A. Hence, its absorption is hampered when there is disorder of the liver.

5.5.5. Storage

Like other fat-soluble vitamins, *e.g.*, A, D and E, vitamin K is also largely stored in the liver and for a prolonged period.

5.5.6. Deficiency Symptoms

Delayed coagulation of blood is most common symptom of vitamin K deficiency although number of platelets are not changed. Profuse blood flow occurs even in minor wounds in deficient condition. Deficiency of vitamin K is more commonly found in children. People suffering from liver and bowel diseases and the alcoholics are more prone to deficiency. Deficiency is observed after long-term use of antibiotics as these destroy intestinal bacteria which synthesize vitamin K. Some common deficiency symptoms are bleeding from nose and gum, anaemia and also osteoporosis and coronary heart disease.

5.5.7. Daily Requirement

Daily requirement of vitamin K in the human body is not precisely assessed but the requirement is extremely small and does not exceed 1 microgram per kg of body weight. Such meagre amount is obtainable from ordinary mixed diet. In case of acute deficiency that may sometimes result after long-term intake of antibiotics, salicylates *etc.* which destroy the intestinal bacteria that synthesize vitamin K, administration of vitamin K may become necessary. Before or after extracting tooth or surgical operation, sometimes vitamin K is advised as preventive measure against bleeding.

5.5.8. Adverse Effects

High accumulation of vitamin K or any of the four fat-soluble vitamins (= a, d, e, k) in the body is harmful. This is because, the excess amount of these vitamins which is not used up remains stored in the body (liver) as they are not easily removed out from the body. Hence, administration of vitamin K, if ever becomes necessary, the dose and interval should be judged with care. Vitamin K in the body if in excess may cause clot formation of blood and thus, thrombosis.

Water-soluble Vitamins

Discussion has so far been made on the fat-soluble vitamins, *i.e.*, vitamins A, D, E and K. Rest part of this Chapter has brought forward discussion on water-soluble vitamins, which comprise vitamins of the B group and the C.

An important point should be kept in mind in regard to water-soluble vitamins. That is, on contrary to the fat-soluble vitamins which are stored in the liver for a considerable period, the water-soluble vitamins are not stored much except in few cases, *e.g.*, vitamin C is stored for some period in the adrenal cortex and vitamin B-12 is stored in the liver. Hence, while the fat-soluble vitamins are drawn upon by the body from the liver when there is any requirement of them in any part, excess of the water-soluble vitamins which is not used is removed out from the body. Removal takes place through urine, faeces and sweat. For this reason, intake of excess amount of water-soluble vitamins which is not required is of no use as the excess is removed out instead of storage.

5.6. B – Vitamins

There are several numbers of vitamins which come under the B group. These are called B vitamins, B group of vitamins or B – complex vitamins. All the individual vitamins that come under the purview of B vitamins have *identical biological role but they differ from one another in respect of chemical structures.*

Most of these B vitamins are *synthesized in the intestine by the action of micro-organisms.* These vitamins function as co-enzymes or prosthetic group of many enzymes. The vitamins of this group have although essential role in the body, some have less requirement, some have still lesser and thus, requirement of some is actually not very clear. There are also some compounds which do not come under the B group of vitamins but they are *related to the B vitamins.*

Among the many forms of the B vitamins, the more common ones may be stated as thiamine, riboflavin, niacin, pyridoxine, pantothenic acid, biotin, folic acid, inositol, cyanocobalamin, para-amino benzoic acid, choline *etc.* These have been discussed in the following:

5.6.1. Vitamin B-1

Among the various members of the B group of vitamins, vitamin B-1 is of special significance. This vitamin is known by other names as thiamine, or aneurine. For its efficacy *to prevent or cure the beriberi disease,* vitamin B1 is also called *anti-beriberi vitamin* (= anti-beriberi factor) and for its beneficial role in the nervous system, this is termed as *anti-neuritic vitamin* (= factor). Symptoms of beriberi had first been induced by Eijkman in chicken by feeding them with rice grains of such type which were made polished. However, recovery from the beriberi had been obtained by Funk when the chicken were fed with only polishings derived from rice. Thiamine in pure form had been obtained by Jansen and Donath in 1926.

5.6.1.1. Functions

Vitamin B-1 has importance in intermediary carbohydrate metabolism. It acts as a co-enzyme in the enzyme reaction which brings about decarboxylation of alpha-keto acids. This has much importance for oxidation of sugar in the brain and the heart.

5.6.1.2. Principal Food Source

The vitamin is abundantly present in the peel (husk) of rice grain (outer layer) and other cereal grains, pulses, underground root vegetables (= carrot, radish, beet, turnip *etc.*), beans, pear, nuts *etc.* Among animal source food, it is rich in egg-yolk, milk, fish, liver, kidney of mutton, beef *etc.*

5.6.1.3. Chemical Aspects

Vitamin B-1 molecule has a pyrimidine ring to which, thiazole is attached. Thiamine is crystalline, whitish and is not heat-stable at higher pH but is much stable to heat at lower pH.

5.6.1.4. Deficiency Symptoms

Under thiamine-deficient condition, normal carbohydrate metabolism is hampered and *pyruvic acid accumulates to a high degree* in different parts of the body. *It results in beriberi disease.* Beriberi may be of (i) dry type, in which nervous system is affected, (ii) wet type, in which cardiac failure occurs and oedema develops and (iii) mixed type, in which symptoms of both dry and wet types are observed. Cerebral beriberi may occur in some persons. Anorexia is also a common symptom. It is seen that in vitamin B-1 deficient condition, ability of birds to fly is lost and in rats, bradycardia develops.

5.6.1.5. Daily Requirement

Daily requirement of vitamin B1 of various age-groups has been stated in Table 17 under the Chapter 9 (Dietetics).

5.6.2. Vitamin B-2

Vitamin B-2 is known as *riboflavin* and other names of it are *lactoflavin, ovoflavin* and *vitamin G*. It is a part of flavonoids and is involved in biological oxidations. Isolation of the vitamin was done by Warburg and Christian in 1932 from yeast.

5.6.2.1. Functions

Riboflavin is essential for growth. It forms co-enzymes, flavin mononucleotide (FMN) and flavin adenine dinucleotide (FAD) to the dehydrogenase enzymes. It is concerned with protein metabolism.

5.6.2.2. Principal Food Source

Milk (lacto-flavin) is a rich source of vitamin B-2. Other food sources are egg, cheese, liver, kidney, soybean, pulse grains, yeast, nuts, leafy vegetables, germinating seeds *etc*.

5.6.2.3. Chemical Aspects

Riboflavin is a nucleoside and occurs as flavin adenine dinucleotide (FAD) or FMN are combined with protein to form flavoprotein. It is made up of a sugar alcohol, ribitol (= alpha-ribose alcohol) and is chemically, 6,7-dimethyl -9-D-l-ribityl-iso-alloxazine.

Riboflavin is light-sensitive, stable in heat but not stable in alkaline condition. The crystals are orange-yellow in colour.

5.6.2.4. Deficiency Symptoms

Under riboflavin deficient condition, growth is retarded. Cheilotic fissure appears at the corners of the lips and lips are also ulcerated. Inflammation of the tongue, *i.e.*, glossitis is a very common symptom. In severe deficiency of riboflavin, eyes are affected, producing symptoms of keratitis, corneal opacities and vascularization. Skin becomes scaly with loss of hair. Severe deficiency forms cataract.

5.6.2.5. Daily Requirement

Stated in Table 17 under the Chapter 9 (Dietetics).

5.6.3. Niacin

This vitamin is nicotinic acid and the nictotinamide which is the amide form of nicotinic acid and is also called niacin-amide. Niacin and niacinamide are known as *anti-pellagra vitamin* and act as co-enzymes (NAD, NADP) to the dehydrogenase enzyme. Hendricks and Smile observed that the heat-stable factors of vitamin B cured pellagra disease. Goldberger first studied that vitamin B enriched food cured pellagra.

5.6.3.1. Functions

Niacin plays part in conversion of *carbohydrate to fat.* In oxidation of tissues, stimulation of nerves, lowering of blood lipid level, preventing pellagra disease and in many other metabolic reactions, this vitamin has important role.

5.6.3.2. Principal Food Sources

Unpolished rice grains and other cereals and yeast, pea, tomato, leafy vegetables, beans and among animal source foods, fish, liver and meat are good sources. Milk is also a good source but cow's milk is a poor source. Niacin is produced in the body from the amino acid, tryptophan also.

5.6.3.3. Chemical Aspects

Niacin is a part of two enzymes in the body, which are NAD (= Co I) and NADP (= Co II). It has a pyridine nucleus with a carboxylic group at the third position and the carboxylic group is replaced by an amide group in nicotinamide.

Niacin is crystalline, readily soluble in alcohol but not completely soluble in water. It can withstand heat.

5.6.3.4. Deficiency Symptoms

Pellagra is a common deficiency symptom of this vitamin but nowadays, this is not a very common disease. Dermatitis, diarrhoea, gastro-intestinal disturbances as well as dementia, mental confusion are other deficiency symptoms of this vitamin.

5.6.3.5. Daily Requirement

Stated in Table 17 under the Chapter on Dietetics.

5.6.4. Pyridoxine

Pyridoxine which is known as vitamin B-6 is a group of pyridine derivatives. This refers to the compounds, *viz*, pyridoxal, pyridoxol and pyridoxamine. These forms are converted into pyridoxal phosphate in the body which is the active form. Pyridoxine forms as the prosthetic group in co-enzymes in transaminase reactions. It was studied by Gyorgi in 1934 that a type of dermatitis could be cured on topical application of an extract made from yeast and pyridoxine had been isolated from the yeast extract.

5.6.4.1. Functions

Pyridoxine acts as a co-enzyme in tyrosine metabolism and tryptophan metabolism. Pyridoxal phosphate is a co-enzyme for the enzymes, *viz.,* aminotransferase, decarboxylase, deaminase and desulphydrase. It is involved in anabolism of lipid and transport of amino acids. Pyridoxine is known to have role in nervous system also.

5.6.4.2. Principal Food Sources

Vitamin B-6, *i.e.,* pyridoxine is rich in germs of cereal grains, leafy vegetables, yeast, nuts and among animal source foods, it is abundant in egg-yolk, meat, liver and kidney. The vitamin is also synthesized by the micro-organisms in the alimentary canal.

5.6.4.3. Chemical Aspects

Pyridoxine is a pyridine derivative and as stated above, the naturally occurring forms of it are pyridoxol, pyridoxal and pyridoxamine while the active form is pyridoxal phosphate.

Pyridoxine is highly soluble in water. It is also heat-stable but destroyed on exposure to light.

5.6.4.4. Deficiency Symptoms

Long-term deficiency of pyridoxine results in lowered level of haemoglobin, some forms of dermatitis (acrodynia), weakness of muscles, degeneration of nerves, mental depression, confusion *etc.* Abdominal pain, vomiting, diarrhoea, convulsion may also occur in deficiency.

5.6.4.5. Daily Requirement

Stated in Table 17 under the Chapter on Dietetics.

5.6.5. Pantothenic Acid

Pantothenic acid, which is also known as vitamin B-3 is a prosthetic group in co-enzyme A and has role in skin, nervous system and gastro-intestinal activity. William and others isolated this vitamin in crystalline form and it is named, pantothenic acid for its widespread occurrence.

5.6.5.1. Functions

As the pantothenic acid acts as prosthetic group in co-enzyme A, it is involved in many metabolic reactions. It takes part in catabolism of fatty acids and is also present in protein-bound form.

5.6.5.2. Principal Food Sources

Pantothenic acid is *widely distributed on Earth and hence, the name panto,* which means all or whole. It is rich in molasses, sweet potato, cabbage, wheat bran, tomato, potato *etc.* and among animal source foods, it is abundant in liver, kidney, egg-yolk and milk. Yeast is also a rich source.

5.6.5.3. Chemical Aspects

Pantothenic acid has an alanine chain in peptide linkage, *i.e.*, peptide of beta-alanine and dihydroxy-dimethyl butyric acid.

Pantothenic acid is heat-stable. The calcium and sodium esters of it are more stable to heat.

5.6.5.4. Deficiency Symptoms

Dermatitis, loss of hair, cornification of skin, degeneration of spinal cord *etc.* are common deficiency symptoms of this vitamin. Symptoms are more perspicuous in animals.

5.6.5.5. Daily Requirement

Stated in Table 17 under the Chapter on Dietetics.

5.6.6. Biotin

Biotin which is co-enzyme-R is termed by some authors as vitamin-H while some authors designate it as vitamin B-7. It is a co-enzyme for carboxylase and is associated with carboxylation reaction. Bateman had observed in 1916 that animals had toxicity when they took high amount of egg-white. Later it was found that a protein, *avidin* isolated from egg-white arrested absorption of biotin.

5.6.6.1. Functions

In carboxylation, biotin is a co-enzyme for carboxylase. It plays part in deamination of some amino acids, like serine, aspartic acid, threonine *etc.* and in synthesis of some fatty acids. It has important role in lipid metabolism.

5.6.6.2. Principal Food Sources

Among animal source foods, biotin is present in high amount in egg-yolk, liver, kidney and muscle. In cooked or unboiled egg, whitish part contains avidin, which antagonizes activity of biotin. Among plant source foods, biotin is present in pea, leafy vegetables, tomato and many fruits. It is synthesized by intestinal bacteria as well.

5.6.6.3. Chemical Aspects

Biotin is a derivative of valeric acid and two five-sided rings are attached to it. Besides, sulphur is also present in it. Biotin is stable in heat as well as in wide range of acidic and alkaline conditions.

5.6.6.4. Deficiency Symptoms

Dermatitis (transient form) results in deficiency of biotin and following that, muscular pain, anorexia and hyperesthesia may occur. Cholesterol in blood may rise in deficiency. Normally, deficiency of biotin does not occur but intake of sulphur drugs and antibiotics may bring about deficiency as the intestinal bacteria which synthesize it are destroyed by sulphur.

5.6.6.5. Daily Requirement

Ordinary mixed diet provides the requirement of biotin. Daily requirement may be 100–250 microgram per day in adults.

5.6.7. Folic Acid

Folic acid (= folate) is also known as pteroylglutamic acid. There are three components of it, which are (i) a pterine base, (ii) para-amino benzoic acid and (iii) glutamic acid. Number of glutamic acid molecules differ. Folic acid is erythropoietic (= formation of red blood cells) vitamin. Reduced form of formylated folic acid is called folinic acid. (Folic acid is sometimes called vitamin-M, though rarely). Mitchell and his co-workers termed it as folic acid because it was isolated from the leaves of spinach and hence, termed as folic acid (folium = leaf). Folic acid has a great role to the women of child bearing age and it has no purpose to administer to the pregnant women. This vitamin prevents neural tube defects in the new borns, which affect their brain and spine.

5.6.7.1. Functions

Folic acid is a co-enzyme in the metabolism of amino acids and in the synthesis of DNA. It is essential in the formation and maturation of red blood cells.

5.6.7.2. Principal Food Sources

Folic acid is abundant in leafy vegetables. Among animal sources foods, it is much present in kidney and liver.

5.6.7.3. Chemical Aspects

Folic acid comprises a class of some chemical compounds of different types. Among these, the pterin derivative which is attached to para-amino benzoic acid and glutamic acid are simplest forms. The reduced form of formylated derivative of folic acid is known as folinic acid.

Folic acid is not much soluble in water but easily soluble in alcohol. Exposure to light for prolonged period destroys its activity.

5.6.7.4. Deficiency Symptoms

In deficiency of folic acid, *megaloblastic anaemia* (megaloblasts are found in blood) is very common and particularly occurs during pregnancy (popularly known as pregnancy-anaemia). Leukopenia (high decrease of white blood cells), inflammation of tongue and other symptoms may also occur. Folic acid has a great role to the women of child bearing age and it has no purpose to administer to the pregnant women. This vitamin prevents neural tube defects in the new borns, which affect their brain and spine.

5.6.7.5. Daily Requirement

Stated in Table 17 under the Chapter 9.

5.6.8. Inositol

Inositol is called *muscle sugar*. It is a sugar-like crystal which is much present in various organs such as muscle, liver, kidney, heart *etc*. It is also abundant in plants. Inositol (also known as inosite or inose) was discovered from plants.

5.6.8.1. Functions

Role of inositol on growth retardation, eye troubles, loss of hair *etc*. has been established in laboratory animals.

5.6.8.2. Principal Food Sources

Among plant source foods, inositol is in high amount in cereal grains and many fruits. (In cereals, it remains as phytic acid). Among animal source foods, it is rich in muscle, brain, milk *etc)*.

5.6.8.3. Chemical Aspects

Chemically, inositol is hexa-hydroxy-cyclo-hexane.

5.6.8.4. Deficiency Symptoms

In experimental rats, fatty liver, alopecia, retarded growth and eye disorder, *i.e.*, spectacle-eye are observed in inositol deficiency. Increase of cholesterol may also take place.

5.6.8.5. Daily Requirement

Not exactly known but 500–800 mg coming from mixed diet is considered satisfactory.

5.6.9 Cyanocobalamin

Cyanocobalamin or vitamin B-12 is also known as *cobamide vitamin*. This vitamin of the B group is stored in the liver for some period. Cyanocobalamin contains in its molecule cobalt cyanide and amino groups. It is also called *non-vegetarian vitamin as it is obtained only from animal source foods, e.g.*, meat, fish, egg and milk. Metabolism of cyanocobalamin is close to folic acid. It takes part in transmethylation reaction and is required for DNA synthesis, myelin formation and is esssential for functioning of nerves and heart. Cyanocobalamin is also called *anti-pernicious anaemia factor*. Smith and Parker isolated it from the liver tissues of animals in 1948.

5.6.9.1. Functions

Cyanocobalamin, *i.e.*, vitamin B-12 is essential for the formation and also maturation of red blood cells. It helps in bone marrow formation and increases white blood cells and blood platelets. The vitamin is necessary for synthesis of DNA, myelin and efficient functioning of nerves.

5.6.9.2. Principal Food Sources

Vitamin B-12 is obtained *only from animal source food, i.e.*, meat, fish, egg and milk and not from any plant. Hence, it is called non-vegetarian vitamin. Liver is the richest source of the vitamin. However, many bacteria which remain present in soil as well as those present in the intestinal tract of human beings and animals can synthesize it. There are reports that some underground root and stem tubers, *e.g.*, potato, carrot, beet, radish *etc.* contain high amount of vitamin B-12. But in fact, the soil-borne bacteria grow on these underground tuber crops and these bacteria actually synthesize vitamin B-12 and not the tubers. Some fungal species, notably, *Streptomyces griseus* can produce vitamin B-12. Human beings cannot synthesize this vitamin and what is synthesized in the alimentary canal is not utilized by them.

5.6.9.3. Chemical Aspects

Molecular structure of cyanocobalamin is composed of 63 carbon atoms. It contains a *cobalt atom in trivalent state.* Cyanocobalamin exists in three forms. In one form, cobalt remains bound to cyanide while in the other two forms, it remains bound to the hydroxyl and nitrite groups respectively. There is also nitrocobalamin, in which nitrogen group is present in place of cyanide group.

Cyanocobalamin is crystalline, red-coloured and water-soluble. Its base is metalloporphyrin and contains cobalt.

5.6.9.4. Deficiency Symptoms

Pernicious anaemia is most common deficiency symptom of vitamin B-12. In the bone marrow, megaloblasts appear in deficiency of vitamin B-12 and folic acid and thereby, brings about macrocytic anaemia. Degenerative lesions in spinal cord, glossitis, mouth ulcers may also occur in deficient condition. Long-term use of metformin drug is known to cause deficiency of vitamin B-12. Absorption of vitamin B-12 in the body however, requires its binding with a chemical known as *intrinsic factor* which is secreted by cells that are lining to the stomach.

5.6.9.5. Daily Requirement

Stated in Table 17 under the Chapter 9.

5.6.10. Para-amino Benzonic Acid

This is an important constituent of folic acid and acts as a co-enzyme in some enzyme system.

5.6.10.1. Functions

Any definite role of this B group of vitamin is not established but its necessity for growth of micro-organisms is known. In rats, it is found to maintain lactation and hair colour is maintained in blackish rats.

5.6.10.2. Principal Food Sources

Among plant source foods, it is rich in pulses, wheat germ *etc.* and among animal source foods, liver and brain are rich sources.

5.6.10.3. Chemical Aspects

The vitamin has structural similarity with sulphonamides which are amides of sulphanilic acid derivatives.

Solubility of the vitamin is less in water and high in alcohol.

5.6.10.4. Deficiency Symptoms

Deficiency symptom of para-amino benzoic acid is not clearly observed in human beings. Graying of hair has however, been reported by its deficiency.

5.6.10.5. Daily Requirement

Daily requirement of the vitamin in human beings is not clearly known. Necessary amount comes to the body from mixed diet.

5.6.11. Choline

In preventing *fatty liver* and damage of kidney, this vitamin is known to have importance.

5.6.11.1. Functions

Choline has lipotropic action. It has role in transmethylation reaction.

5.6.11.2. Principal Food Sources

Among plant source foods, choline is rich in rice bran, wheat germ, pea, beans *etc.* while liver, kidney and egg-yolk are rich sources among animal source foods.

5.6.11.3. Chemical Aspects

Chemically, choline is trimethyl-hydroxyethyl-ammonium hydroxide.

5.6.11.4. Deficiency Symptoms

Deficiency symptoms of choline is not very clear as it is naturally synthesized in the body of human beings.

5.7. Vitamin P

An extract was made by Szent-Gyorgy in 1936 from paprika (= chilli) and lime juice and it was observed that in addition to vitamin C, the extract was essential to prevent capillary fragility in human beings. He proposed the name vitamin P to the factor as the extract was made from paprika (using the letter P of paprika). However, later the Joint Committee on nomenclature of the American Society of Biological Chemists and American Institute of Nutrition (1950) proposed that the term, vitamin P should be dropped for this factor. Instead of that, for these group of compounds, the term bioflavonoids should be appropriate and hence, referred to as bioflavonoids.

5.7.1. Functions

It has been confirmed that vitamin P, *i.e.*, bioflavonoids have their value in preventing capillary fragility along with vitamin C. They act in the same way as vitamin C.

5.7.2. Principal Food Sources

These bioflavonoids are mostly present along with vitamin C. Thus, fruits that are rich source of vitamin C are usually good sources of these flavonoids.

5.7.3. Chemical Aspects

Vitamin P is nowadays universally considered as bioflavonoids. Among these, rutin, esculin, citrin and quercitrin *etc.* are of importance. Rutin is chemically, penta-hydroxy flavone-3-rutinoside. Esculin is dihydroxy-coumarin-6-glucoside. Citrin, which is obtained from lemon peel is a mixture of two glucoside, *i.e.*, hespiridin and eriodictin.

5.7.4. Deficiency Symptoms

Scarborough (1940) studied that deficiency of vitamin P, *i.e.*, flavonoids were responsible for decreased capillary resistance bringing about petechial bleeding with pain on shoulders and legs, lassitude, fatigue *etc.*

5.7.5. Daily Requirement

Not clearly known. However, bioflavonoids are efficacious at a dose of 800 mg a day orally in curing lowered capillary resistance in allergic reaction. Intramuscular administration may produce toxicity.

5.8. Vitamin C

Vitamin-C, other name of which is *ascorbic acid* is nowadays considered as a chemical compound of great worth and is not a vitamin only. It is an *excellent anti-oxidant* that does not produce any side-effects and has tremendous therapeutic properties. For *synthesis of the collagen protein* in the body which is of great value, vitamin-C is indispensable. At least eight enzyme systems in the body act, using vitamin-C as the co-factor. The vitamin plays great role in boosting the immune system, digestion of iron, increasing white blood cells, antibodies *etc.* Even in plants, vitamin-C plays great role in many physiological mechanisms and is recognized as a plant hormone. The property of vitamin-C to release ethylene gas inside plant body has been reported.

5.8.1. Discovery

Scurvy was once a dreaded disease in the whole world and the necessity of preventing and curing this disease led to the remarkable discovery of vitamin-C. The disease was known even 2500 years ago but its scientific study was actually initiated in the 18th century. It is said that the Portuguese sailor, Vasco-de-Gama sailed for India in 1497 with a crew of 160 sailors but when the ship reached India, 100 sailors died by scurvy in the sea.

James Lind, the surgeon of the British Navy took up extensive studies on the disease. It came to his notice that in the ship, the sailors had to take only stored food in their voyage for months together and there might be some defect in the food which had to be stored for such long period. He observed that scurvy disease could be greatly averted among the sailors by feeding them with juice of sour lime fruit (a type of citrus). He stated his observation in 1753 in a journal but his article had not received heedful consideration. This is because, the scientists then were of opinion that scurvy disease had nothing to do with food and it could be prevented by physical exercise and behavioural change. However, subsequently, Captain Cook greatly averted scurvy disease of sailors by giving them raw (= uncooked) onion to them in the Australian voyage (1772-1775).

Much later, the scientists, Axel Holst and Theodor Frolich in 1907 came to know that when the food materials were stored for long time in the ship *under ordinary (existing) temperature condition, their active ingredient which was of health benefit was destroyed* (at that time, there was no refrigeration in ship). As a result, the sailors were afflicted with scurvy disease. It was observed by the scientists at that time that the so-called active ingredient(s) which was responsible to keep scurvy at bay was *acidic in nature*. It was also observed by them that if only the acidic substance on suitably extracting from different food-stuffs could be fed to the sailors or other persons, instead of food, scurvy disease had been greatly prevented in them.

The disease, scurvy *was then known as scorbutic disease* and as the extracted material from different food-stuffs which arrested the disease had been *acidic in nature, the extracted material was named as anti-scorbutic acid, i.e.*, the acid which prevented scurvy. Later, the term, anti-scorbutic acid being a big word was shortened by deleting some letters from it and it was termed as *ascorbic acid*. Now the term, ascorbic acid is prevalent, which is the other name of vitamin C.

Early works on vitamin-C had largely been undertaken during 1915 to 1925 by Zilva and it was observed that the extracts of vitamin-C had *strong chemical reduction activity*. Later, extracts made from citrus fruits and adrenal gland, Szent-Gyorgy (1928) also observed reducing properties of the isolates. He termed it hexuronic acid but it was identical to ascorbic acid. Isolation of vitamin-C in crystalline form was later made by Waugh and King (1932) from lemon juice. Synthesis of vitamin-C had been done (1933) independently by Haworth and others in U.K. and Reichstein and others in Switzerland.

At present, scurvy is not considered to be a serious disease. This is because, it is easily averted by taking vitamin-C and if the disease appears, it is cured by administration of vitamin-C which is specific remedy to the disease. Instead of taking vitamin-C as a medicine, foods rich in this vitamin may also be taken.

5.8.2. Functions

Vitamin C has plentiful physiological functions and some have been stated in the following:

(i) *Synthesis of collagen protein* – Collagen is an elongated, fibril protein which is triple helical. The name is derived from the Greek word, *kolla* which means glue. This protein connects and supports body tissues, *e.g.*, skin, bone, muscle, ligaments, teeth *etc.* Collagen has also tendons, therapeutic properties in heart diseases, arthritis, varicose vein *etc.* This protein is *most plentiful of all proteins in human beings* and constitutes 25–35 per cent of all proteins in the body and 75 per cent of proteins in skin. It maintains elasticity of skin. Under collagen deficient conditon, a number of ailments take place in the body. *Vitamin-C indispensable in synthesis of collagen.* It brings about change of the amino acid, proline into hydroxyproline and the amino acid, lysine into hydroxylysine. Synthesis of collagen by the action of vitamin-C occurs in several steps.

(ii) *Reducing property* – In many chemical reactions, vitamin-C is a potent reducing agent. By reducing the ferric form of iron to the ferrous form, it *promotes absorption of non-haem part* from the gut and helps in stabilizing iron-binding protein. Vitamin-C and adenosine triphosphate (ATP) play part for intake of plasma iron in the tissues as ferritin as well.

(iii) *Anti-oxidant effect* – *Vitamin-C is a powerful anti-oxidant* and thus, removes free-redicals (= reactive oxygen species) from the body. Thereby, it protects the body against atherosclerosis, hypertension, cardiovascular diseases, ageing, inflammation of various types and also cancerous growth.

(iv) *Tyrosine metabolism* – It is found that by feeding tyrosine to rats afflicted with scurvy, they excrete homogentisic acid, para-hydroxy phenyl lactic acid and para-hydroxy phenyl pyruvic acid. But these excretions do not occur when vitamin C is fed to them.

(v) *Transport of electrons* – In many neurochemical reactions related to transport of electrons, vitamin-C takes part.

(vi) *Hydroxylation* – In hydroxylation of deoxycortico-sterone, hydroxylation of 5-hydroxy-tryptophan from the amino acid, tryptophan, vitamin-C plays part.

(vii) *Co-enzyme* – Vitamin-C acts as a co-enzyme for at least eight enzymes in the body.

(viii) *Immunity* – Boosting up immunity in the body is an important property of vitamin-C.

(ix) *Synthesis of cortical hormone* – Vitamin-C has role in cortical hormone synthesis.

(x) *Bone health* – Vitamin-C has importance in proper formation of bone matrix.

5.8.3. Principal Food Sources

Fruits of many plants are very rich sources of vitamin-C. Germinated seeds of pulses also contain high amount.

5.8.4. Chemical Aspects

Vitamin-C is a dibasic acid with an enediol group which is built into a five-membered lactone ring (heterocyclic). It exists naturally in *laevo form which is physiologically active*. The other form, *i.e.*, dextro form is physiologically inactive. Vitamin-C has in fact, a number of vitamers which have vitamin-C activity. It is an electron donor for many enzymes.

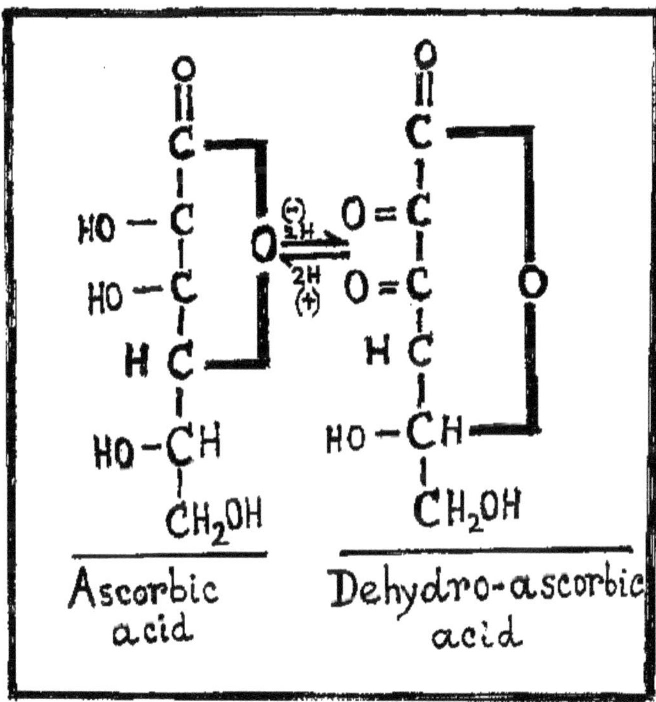

Figure 16. Structures of Ascorbic Acid (reduced form) and Dehydro-ascorbic Acid (oxidized form).

On oxidation, vitamin-C gives rise to *dehydro-ascorbic acid*. Both the dehydro-ascorbic acid which is the oxidized form and the ascorbic acid, which is the reduced form are physiologically active. Reaction of ascorbic acid and dehydro-ascorbic acid is reversible.

Vitamin-C is crystalline, white and highly soluble in water but not soluble in fat, oil or fat-solvents. Its activity is lost in heat. In presence of air, *i.e.*, oxygen, it is readily oxidized at 100°C. It makes chemical reduction of methylene blue, 2, 6-dichlorophenol indophenol and turns those to red colour. Reduction is also done to silver nitrate and ferricyanide solution.

In acidic solution, vitamin-C remains active and the activity is lost in alkaline medium. Copper salts inactivate activity of vitamin-C and hence, cooking food in copper utensils destroys activity in the food. Cooking of food in presence of air destroys vitamin-C activity due to *oxidase enzyme* but cooking done in closed chamber (in absence of air) saves the food from loss of vitamin-C. Activity of vitamin-C is much lost if heating is done slowly rather than when cooking is done quickly on heat.

5.8.5. Absorption

Vitamin-C is readily absorbed from the intestine. It is absorbed by both active transport and diffusion. The two transporters for absorption are, sodium-ascorbate co-transporters (SVCTs) and hexose transporters (GLUTs). Absorption of dehydro-ascorbic acid takes place in higher rate than ascorbic acid but the amount of the former form is lower in the plasma and tissues as the cells readily reduce it to ascorbic acid.

5.8.6. Deficiency Symptoms

Human beings should not have been suffered from deficiency of vitamin-C, because primitive people were able to synthesize vitamin-C in their own body to meet the requirement. However, with passage of time, the *gene responsible for synthesis of vitamin-C in the body had been mutated and had become non-functional in human beings* as well as in many other animals like apes, many species of bats, fishes, capibara, birds, guineapig *etc.* As a matter of fact, the last of the four steps that are necessary for vitamin-C synthesis is not possible to be performed now by these animals including human beings. Hence, it is essential now for the human beings and these animals to obtain vitamin-C from external sources (food) to meet their requirement in the body. But many other animals are able to synthesize vitamin-C in their body.

A number of symptoms appear in human beings due to deficiency of vitamin-C and the more common ones may be enlisted as follows:

(i) Scurvy, *i.e.*, scorbutic disease is, by far, most dreadful. This causes bleeding from gums, eyes and skin. Brown patches appear on skin and especially on thighs. Body gets fever and if not timely treated, it becomes fatal and life-threatening.

(ii) Growth is retarded in children.

(iii) Bones become porous and brittle.

(iv) Gingivitis (gum inflammation) is very common.

(v) Elasticity of skin is hampered and hence, skin becomes dry, scaly and wrinkled. Hairs become split.

(vi) Microcytic anaemia may occur.

(vii) Blood clotting (= coagulation) is delayed.

(viii) Healing of wound is impaired.

(ix) Swelling and pain occur in joints.
(x) Carbohydrate metabolism is disturbed.
(xi) Neuromascular coordination is disturbed.
(xii) Reproductive failure may take place both in men and women.
(xiii) Blood sugar level may rise.
(xiv) Long-term deficiency of vitamin-C brings about atherosclerosis and hypertension.

It should be noted that scurvy is not a serious disease now, because vitamin-C being a *specific antidote* of it, the disease is easily cured at its early stage of appearance by administering vitamin-C.

5.8.7. Daily Requirement

Daily requirement of vitamin-C for persons of various age-groups has been stated in Table 17 under the Chapter 9. Requirement however, varies according to authors. For therapeutic purpose, mega dose of vitamin-C is however, advised by many health specialists.

5.9. Carotene, Vitamins of the B Group and Vitamin C Content of some Common Food-stuffs

Carotene, vitamins of the B group and C content as stated by Gopalan *et al.* in their book, "Nutritive Value of Indian Foods." (2018), revised by Narasinga Rao *etal.* (published by the National Institute of Nutrition, Hyderabad, under the Indian Council of Medical Research, New Delhi) have been enlisted in Table 14. The amount for some selected food-stuffs have only been presented. For the amount present in other food-stuffs, the original title may be consulted.

Table 14. Carotene, Vitamins of the B Group and C Content in some Food-stuffs under different Groups

Food-stuffs	*Carotene, Vitamins B and C*								
	**A*	*B*	*C*	*D*	*E*	*F*	*G*	*H*	*I*
				****I.**					
a	132	0.33	0.25	2.3	–	14.7	45.5	0	–
b	10	0.47	0.20	5.4	–	–	–	–	–
c	47	0.37	0.13	3.1	0.21	14.0	20.0	0	–
d	9	0.27	0.12	4.0	–	–	–	0	–
e	–	0.21	0.05	3.8	0.24	8.9	11.0	0	–
f	2	0.21	0.16	3.9	–	–	–	0	77
g	0	0.06	0.06	1.9	–	4.1	8.0	0	–
h	–	2.7	0.48	–	–	–	–	0	–
i	29	0.49	0.17	4.3	–	12.1	35.8	0	–
j	25	0.12	0.07	2.4	–	–	0	0	–

Food-stuffs	Carotene, Vitamins B and C								
	*A	B	C	D	E	F	G	H	I
II.									
a	129	0.48	0.18	2.4	–	32.0	147.5	1	–
b	38	0.42	0.20	2.0	–	24.0	132.0	0	206
c	49	0.47	0.21	2.4	–	24.5	140.0	0	–
d	132	0.45	0.19	2.9	0.54	19.0	103.0	0	183
III.									
a	120	0.06	0.09	0.4	–	13.3	23.0	124	120
b	7560	0.08	0.21	2.3	–	23.5	93.9	4	–
c	990	0.09	0.13	0.50	–	–	–	10	170
d	5580	0.03	0.26	0.5	–	51.0	123.0	28	–
IV.									
a	0	0.04	0.09	0.4	–	–	–	10	242
b	0	0.08	0.01	0.4	–	1.5	6.0	11	–
c	6	0.08	0.04	0.7	–	–	–	24	–
d	0	0.04	0.04	0.5	–	–	–	43	135
V.									
a	126	0.07	0.09	0.5	–	–	–	88	–
b	0	0.03	0.01	0.2	–	–	–	0	–
c	30	0.04	0.10	1.0	–	–	–	56	127
d	52	0.07	0.10	0.6	–	25.3	105.1	13	–
e	50	0.06	0.04	0.5	–	3.0	13.0	2	136
VI.									
a	0	0.24	0.57	4.4	–	–	–	0	–
b	60	0.63	0.19	1.2	–	–	–	0	–
c	0	0.05	0.10	0.8	–	11.7	12.5	1	–
d	144	0.67	0.28	2.3	–	–	–	–	–
e	6	0.45	0.40	1.0	–	–	–	0	–
VII.									
a	175	0.19	0.39	0.9	–	6.0	29.0	111	–
b	30	0.03	0	2.3	–	10.0	18.0	0	–
VIII.									
a	9	0.03	0.01	0.2	–	–	–	600	256
b	0	–	–	0	–	–	–	1	321
c	55	0.13	0.03	1.1	–	–	–	8	–
d	78	0.05	0.08	0.5	–	–	–	7	1
e	3	0.04	0.03	0.2	–	–	–	1	–
f	0	0.03	0.03	0.4	–	–	–	212	–

Food-stuffs	Carotene, Vitamins B and C								
	*A	B	C	D	E	F	G	H	I
g	0	0.02	0.06	0.4	–	–	–	31	–
h	2743	0.08	0.09	0.9	–	–	–	16	–
i	666	0.04	0.25	0.2	–	–	–	57	–
IX.									
a	–	–	–	0.7	–	–	–	10	349
b	–	–	–	2.8	–	–	–	24	1364
c	–	–	–	0.8	–	–	–	–	611
d	–	0.05	0.07	0.7	–	–	–	22	819
X.									
a	405	0.12	0.26	0.2	–	80	80	–	–
b	420	0.10	0.40	0.1	–	70.3	78.3	0	–
c	–	–	0.14	–	–	3.2	6.8	–	–
d	9	0.18	0.14	6.8	–	1.0	5.8	–	–
XI.									
a	48	0.04	0.10	0.1	–	3.3	5.6	1	–
b	53	0.05	0.19	0.1	–	5.6	8.5	2	–
c	55	0.05	0.04	0.3	–	0.7	1.3	1	–
d	41	0.02	0.02	–	–	1.3	–	3	–
XII.									
a	960	–	–	–	–	–	–	–	–
b	600	–	–	–	–	–	–	–	–
c	270	–	–	–	–	–	–	–	–

*A: Carotene - microgram - per 100 gram of edible matter. B: Thiamine - milligram - per 100 gram of edible matter. C: Riboflavin - milligram - per 100 gram of edible matter. D: Niacin - milligram - per 100 gram of edible matter. E: Total B6 - milligram - per 100 gram of edible matter. F: Folic acid (free) - microgram - per 100 gram of edible matter. G: Folic acid (total) - microgram - per 100 gram of edible matter. H: Vitamin C - milligram - per 100 gram of edible matter. I: Choline - milligram - per 100 gram of edible matter.

****I: Cereal grains** – a – Bajra. b – Barley. c – Jowar. d – Rice, parboiled, handpounded. e – Rice, parboiled, milled. f – Rice, raw, handpounded. g – Rice, raw, milled. h – Rice, bran. i – Wheat flour, whole. j – Wheat flour, refined. **II: Pulses** – a – Bengal gram, daal. b – Black gram, daal. c – Green gram, daal. d – Red gram, daal. **III: Leafy vegetables** – a – Cabbage. b – Curry leaves. c – Lettuce. d – Spinach. **IV: Roots and tubers** – a – Beet root. b – Onion, big. c – Sweet potato. d – Turnip. **V: Other vegetables** – a – Bitter gourd. b – Bottle gourd. c – Cauliflower. d – Okra. e – Pumpkin. **VI: Nuts** – a – Almond. b – Cashewnut. c – Coconut, fresh. d – Pistachio nut. e – Walnut. **VII: Spices and condiments** – a – Chilli, green. b – Turmeric. **VIII: Fruits** – a – Emblica. b – Apple. c – Bael. d – Banana, ripe. e – Grape, blue variety. f – Guava. g – Litchi. h – Mango, ripe. i – Papaya, ripe. **IX: Fishes** – a – Bhekti, fresh. b - Hilsa. c – Katla. d – Rohu. **X: Meat, fowl and eggs** – a – Egg. b – Egg, hen. c – Fowl. d – Mutton. **XI: Milk** – a – Milk, buffalo's b – Milk, cow's. c – Milk, goat's. d – Milk, human. **XII: Fat** – a – Butter. b – Ghee, cow. c – Ghee, buffalo.

Chapter 6

Minerals

6.1. General Aspects

A wide variety of mineral elements are indispensable in the body to maintain health and to carry on the daily activities effectively. Like protein, carbohydrate, fat, vitamins and water, minerals also *constitute one of the six components of nutrients in the human body.*

Among the large number of mineral elements that are taken up by the human body, some are required in relatively high amount, some in low amount and some in very low or in extremely low amount, *i.e.*, in modicum. But whatever amount that is taken up for a given mineral element according to its requirement is considered to be *essential, i.e.*, indispensable. It is observed that health hazard may occur even if there is little deficiency of an element.

Each mineral element has in fact, a set of physiological functions in the body and any dearth or sub-optimal provision of that brings about unhealthful situation. It should be remembered that *starvation from mineral elements results in death of a person much earlier than in case of food starvation.*

Most of the mineral elements that are taken up by the body are usually of *low atomic weight.* It may be pointed out that deficiency symptoms of some mineral elements are not detectable at early stage. This does not however, indicate that the person has no deficiency of those elements. In fact, deficiency symptoms of some elements appear sometime later, which are referred to as *delayed symptoms of mineral deficiency.*

There are many animals which are at times found to meet up the requirement of some mineral elements by consuming some uncommon or odd substances. For example, some street cows are sometimes found to eat even human excreta and they do so when there is severe deficiency of phosphorus in their body. In the human beings also, such peculiar habit of eating odd substances is not uncommon. For example, it is a practice among many rural women in India to eat burnt clay or ash during their pregnancy. *This is called pica and is even advised by the elderly women* in the family to take during pregnancy and lactation. Unknowingly, this is practised by them to get into the body higher amount of some minerals, necessity of which is felt. Eating uncommon materials has also to be done by military personnels during war and by the common people when there is prolonged famine or flood when food is scarcely available.

Minerals are *inorganic substances* and thus, are not produced in the body. Hence, they are to be obtained from *external source, i.e.*, through food and drink. However, sometimes the body gets some minerals(s) from within the body. For example, when a red blood cell (corpuscle) is broken down, the iron that is present in the haemoglobin inside the red blood cell is made free. This free form of iron which is liberated by the broken haemoglobin is re-used in the body to form new haemoglobin.

In order to secure all the required minerals in proper amount, the body must depend on taking staple foods and drinks regularly and failing to do so will evidently, act upon health. It should be noted that all the mineral elements that are required by the human beings exist on Earth as salt except sodium, chlorine and iodine that are obtainable from sea-water. Human beings get those as salts or through other plants or animals in food-chain.

It should be remembered that the amount of any mineral that the body takes up is *not very rigid*. It widely varies according to physical condition of people, their age, gender, pregnancy, lactation, health and many other internal factors. External factors also have great influence and *eco-environmental condition* of the localities where people live also play part. For example, people living in an area having soil rich in iron are apt to take up high amount of iron in their body. Those living in areas having soil or water containing high amount of lime tend to deposit high calcium in their body. Sometimes it is observed that the body takes up fairly high amount of one or more mineral salts, particularly in areas where these are available in plenty. The amount that is taken up is much higher than what is required and is called *luxury consumption*.

To seek for the answer as to how the various mineral elements are utilized in the body, it should be stated that they have indispensable role in many metabolic reactions and that they act as metallic co-factors in many enzyme systems. Apart from these, some of the minerals act as *essential constituents* of some organs or compounds in the body and no doubt, formation of the organs or compounds will be hampered in ther absence. Like vitamins, minerals do not however, give rise to any energy in the body directly.

Mineral elements comprise inorganic substances. That is, when any of the mineral element barring carbon is burnt or even ashed, *it remains the same* without altering the physical and chemical properties and is not destroyed, *i.e.*, lost. That is, the total mineral matter content of the human body, which constitutes 4 per cent of body weight and which are taken up from the Mother Earth ultimately *go back to the Earth in the form of minerals*. In other words, after death, if the body is buried or even burnt to ashes, mineral matters are not changed although all other organic materials are lost, better to say, transformed into other matters through formation of gases. Thus, what the body of a person takes up from the Mother Earth in lifetime is *returned to Her*. The anatomical scientists having philosophic mind therefore, regard the mineral elements of a living body as *aatmaa* (= spiritual being) in the body following the doctrines of the holy Geeta of the Hindu religion which propounds that *aatmaa is non-destructible and even fire cannot burn it.*

Accumulation of mineral salts in the body of the human beings as also in all animals has in fact, been a trend since pre-historic times. C. C. Chatterjee (2004) in his book (Human Physiology, Volume 1, Medical Allied Agency, Kolkata - 700 009) has stated that the mineral salts which were in existence in the primitive living being which was born in sea-water still remain in the plasma of bood of today's human beings and animals. Physiological functions of the mineral elements also remain the same as with the earliest form of life. Thus, the human beings and animals could be said to have been copying their earliest ancestor in respect of mineral metabolism in their body.

6.1.1. Definition

Minerals are naturally occurring chemical elements that exist on Earth as inorganic elements or salts or components of organic compounds. They are solid in form except mercury which is liquid. Each mineral element has a set of characteristic *physical and chemical properties*.

A number of mineral elements are vitally necessary to the human beings, animals and plants for a number of physiological functions to perform. Some are also constituents of some compounds in their body. Carbon is although a material obtained from the mines, it is not considered in the group of minerals required by the body. It is essentially required to form any organic compound and is decomposed into oxide form. A number of mineral elements are required by the human body among which, some have indispensable requirements. The mineral elements that are required by the human beings vary in their amount and some like calcium, phosphorus, magnesium, sodium, potassium, chloride *etc.* are usually required in relatively high amount.

6.1.2. Classification

Quite a large number of mineral elements are taken up by the body daily through foods, drinks and salt. These are broadly classified into two groups, which are termed as *essential and non-essential mineral elements.*

Essential elements are those, which are unavoidable to the body for maintenance of health and effectively carrying on daily activities. Deprivation of any of these essential elements may seriously act upon health. Non-essential elements though have not essential requirement but their intake are also of much help to the body in many ways.

The amount of the essential mineral elements that are required or taken up by the body greatly varies according to the concerned element. It is known that some elements are required in higher amount than other elements. For example, calcium constitutes 1.5 per cent of body weight, phosphorus constitutes 1 per cent and potassium, sodium, chlorine, sulphur and magnesium constitute 0.85 per cent of body weight in human beings.

Thus, the various essential mineral elements are conveniently made into two groups in accordance with the *relative amount* of them that are required in the body. These are termed as follows:

(1) *Macro-mineral elements*, also referred to as major- or primary-mineral elements.

(2) *Micro-mineral elements*, also referred to as minor- or secondary-mineral elements. These are also termed as *trace elements*.

Two important aspects are to be critically considered in respect of the above classification. These are as follows:

(i) The classification is based on *relative amount of the mineral elements only* that are actually taken up by the human body. This signifies that if an element is required by the body in relatively high amount, it should be placed under the category of macro-elements and likewise, if an element is required in relatively low amount, it should be recognized as a micro-element. There is actually *non specificity as regard the amount of a mineral* that should be considered to group the minerals into the major and minor classes. Some authors consider that a mineral which is required in more than 100 mg per day should better be considered as a major element and if the requirement is less than 100 mg per day, it should be considered to be a minor element.

The above numerical criterion has not however, been universally accepted. The nutritional scientists have in fact, opined that the relative level of the mineral that is taken up should only be considered although the term, relative has vagueness.

(ii) The mineral elements belonging to each group are of *equal physiological importance*. It should never be thought that the minor elements are of lesser or secondary importance than the major ones, because the term, minor is prefixed to them. The term, minor indicates that they are required in relatively low amount than those which are regarded as major minerals. Minor, *i.e.*, micro-minerals are required in very small amount and in traces.

Hence, they are usually called *trace-elements*. Some of them are required in extremely low amount and these are recognized as *ultra-trace elements.* It should be remembered that even if a modicum of an element belonging to the ultra-trace group is required, that has physiological significance which is *no less than* any other macro-element or micro-element.

List of Macro- and Micro-mineral Elements

Of about 25–29 mineral elements that are taken up and utilized by the human body for nutrition, which ones have to be regarded as macro-elements and which ones as micro-elements are in fact, difficult to be assessed with exactness. Although some elements are universally taken up by the human beings in relatively high and relatively low amount, variations are also noticed under certain conditions. It may be stated that such variations occur for a number of factors and some may be mentioned as *age and gender of people, pregnancy, lactation, physical and mental activities, health, eco-environmental conditions of living etc*. An example may be given of the people who inhabit in very high altitude in the mountains. Due to oxygen tension in very high altitudes, people living there have to synthesize much higher amount of haemoglobin in their body and this necessitates high intake of iron (People in these localities are found to develop even *red spots* on their skin due to high haemoglobin content of blood). On the other hand, in the sea-shore areas particularly in the tropical countries, oxygen concentration in the air is much high and people living in these parts have opposite trend. Classifying mineral elements into major- or minor-groups also depend on the authors, *i.e.*, while some authors consider some elements into the micro-group, others prefer to group them as micro-elements.

In a general manner, the following elements are usually considered as macro-mineral elements.

Calcium, phosphorus, potassium, sodium, magnesium, chlorine, sulphur *etc*.

On contrary, the elements listed in the following are usually considered as micro-mineral elements or are also otherwise termed as *trace-elements*.

Iron, zinc, iodine, copper, selenium, manganese, chromium, cobalt, molybdenum, silicon, nickel, aluminium, germanium, arsenic *etc*.

It should be noted that all these trace-elements may not be genuinely necessary under all conditions. Further, some of the trace-elements may though be required but in extremely low amount only and hence, these are termed as *ultra-trace mineral elements*.

Despite the fact that trace- or ultra-trace elements have their necessity in the body, it may be mentioned that definite role, physiological actions, quantity required and symptoms of deficiency of them have although been established in animals, these have not been precisely assessed in case of human beings.

6.1.3. Physiological Functions

Mineral elements have plentiful physiological functions in the body of human beings. Some of them are also essential constituents, *i.e.*, ingredients of some organs in the body and therefore, formation of the concerned organs will not take place or may be deformed in absence or shortage of concerned mineral elements. Some elements are *constituents* of a number of organic compounds. Some act as *metallic co-factors* in many enzyme systems. Maintaining *electrolytic balance* in and outside cells is an important function of the mineral elements. *Transportation,* maintaining *osmotic pressure*, controlling plasma *glucose* and in many other physiological processes, the mineral elements are known to be indispensable. Few examples are stated in the following:

Minerals like calcium phosphorus and magnesium are constituents of bones and teeth. Phosphorus is also a constituent of bones and teeth and soft tissues like muscles, liver *etc.* Iron is a component of haemoglobin.

Iodine is an essential necessity for production of thyroxin, zinc in insulin, sulphur in thiamine and cobalt in vitamin B-12. Phospholipids contain phosphorus which are components of cell membranes and phosphorus is s constituent of nucleic acids.

Many enzymes require minerals for their activity, *e.g.*, cytochrome in mitochondria. Zinc is a metallic co-factor of plentiful enzymes. Zinc and copper are part of anti-oxidative enzymes. Magnesium takes part in many enzyme systems. Some enzymes are activated in presence of phosphorus. Copper takes part in formation of red blood cells. Chromium has its necessity for glucose and lipid metabolism.

To maintain proper osmotic pressure of body fluids, electrolytes such as sodium and potassium are essential. Minerals regulate permeability of cell membrane. They are necessary for secretion of glands. Blood volume is regulated of them. Calcium, potassium and sodium have necessity to maintain rhythm of heart beat. In clotting of blood, calcium and phosphorus have important role. Sulphur is a constituent of three amino acids (= methionine, cysteine and cystine) which have importance in the health of skin, hair, nails and liver. Zinc promotes healing of wounds. Neutral reaction (pH) of blood is also maintained by many minerals.

6.1.4. Absorption

Absorption of mineral elements largely takes place in the small intestine. Some minerals may however, be absorbed in the large intestine also but only little amount.

As regard calcium, absorption of it takes place in presence of vitamin-D, lactose and protein. An active transport mechanism is involved in the case of calcium. Absorption of iron is associated with release from the iron-phosphate-protein compound, *ferritin* which stores iron. Phosphate is absorbed all along the small intestine but to a greater part from the ileum.

Absorption of mineral salts in regulated by a number of physical and chemical factors. However, necessity of a mineral in the body is an important determining factor in regard to the amount and quickness of absorption. For example. calcium absorption is higher when there is deficiency of it. Similarly, iron and copper absorptions are higher in anaemic condition in order to raise the level of haemoglobin in the blood.

It should be remembered that some mineral elements facilitate absorption of other mineral elements while some may slow down or inhibit absorption of other mineral elements, *i.e.*, they are *antagonistic* to them. Of great consequence is some chemical compounds which greatly reduce or inhibit absorption of some mineral elements. These are referred to as *anti-nutritional factors*, which have been dealt in Chapter 8.

6.1.5. Storage

Almost all the minerals that are taken up by the human body are stored for short or long period but some may be stored for a great length of period. Some vitamins like vitamins A, D, E, K besides B-12 and C are also stored in the body for a considerable period for enzymatic activity and the mineral elements are stored for metabolism and structure of organs.

The minerals, *viz.*, calcium, phosphorus and magnesium are stored in the bones and teeth. These organs store about 99 per cent of body's calcium and more than 65 per cent of body's phosphorus content. Storage of iron in the body is done mainly in the liver but is also stored in bone-marrow. Spleen also stores iron to a considerable amount.

The specific influences of some of the important macro- and micro-mineral elements on human health have been discussed under the following heads.

6.2. Calcium

Calcium is most common of the mineral elements as *it maintains overall health*. The element constitutes 1.5 per cent of body weight.

In human anatomy, calcium is stated to be "the bone and tooth material". This is because, this mineral forms 99 per cent of bones and teeth, *i.e.*, the body uses 99 per cent of calcium to form bones and teeth. In the bones and teeth, calcium is present *mostly as calcium phosphate and partly as calcium carbonate*. Calcium and phosphate ions are in fact, *complementary to each other*. That is, without phosphate, calcium cannot be retained in the body nor it cannot function properly and *vice versa*. Large number of minute crystal-like substances are located in the bones. These are called *hydroxyapatites*, in which high amount of calcium is present.

Apart from bones and teeth, calcium has its role in other body systems. It also plays a vital role in intracellular signalling to bring about integration and regulation of metabolic processes. The body needs the mineral all throughout the life and also in old age when the calcium level tends to decrease.

6.2.1. Functions

Calcium has plenteous functions in the body and some important ones are stated below in short.

(i) As stated under the Heading 6.2, formation and development of bones and teeth cannot be conceived of without adequate calcium in the body.

(ii) It has a great role in contraction and relaxation of all types of muscles.

(iii) In the nervous systems, the mineral helps in maintaining a normal response to stimuli.

(iv) Calcium maintains rhythm in heart beat. (Role of calcium ions in cardiac contraction was first discovered by Sydney Ringer in 1883, however, accidentally).

(v) Activation, regulation or inhibition of many enzymes are done by this mineral element.

(vi) Calcium ions have necessity in coagulation of blood.

(vii) The mineral supports immune system.

(viii) Blood pressure in regulated by calcium ions.

(ix) The mineral element is known to prevent colo-rectal cancer.

(x) It is an integral part of biomembranes.

(xi) Calcium acts as a link between exciation and secretion of endocrine as well as exocrine glands.

(xii) In preventing obesity and relieving insomnia, the mineral has adequate role.

(xiii) Calcium ions are highly efficacious in controlling allergic reactions and skin eruptions.

6.2.2. Principal Food Sources

Milk is by far, the most important and potential source of calcium. Among other animal source foods, oysters, cheese and particularly, sea-fishes, notably salmon, sardine *etc.* are worth mentioning. Among plant source food-stuffs, almond, collards, spinach, kale, molasses, broccoli, peas and other legumes, drumstick, oranges *etc.* are rich sources.

6.2.3. Chemical Aspects

Calcium which comes under Group II – A of Periodic Table has atomic weight 40.08, atomic number 20 and +2 as valence.

6.2.4. Absorption

The mineral is largely absorbed in the duodenum. However, some amount is absorbed from other parts of small intestine also. Absorption of calcium is a difficult process as it forms insoluble phosphate. Calcium absorption has a threshold level and it depends on calcitriol. Lower pH level favours absorption. Adequate level of some amino acids also make higher absorption of calcium. Absorption of calcium is much inhibited by anti-nutrient factors as stated in the Chapter 8. Steroids and prolonged use of antacids also lower the rate of calcium absorption in the body. Microtears in the bones are however, necessary for the calcium to enter into the bones and this is facilitated by some weight-bearing physical exercises. such as running, jogging, jumping *etc.*

6.2.5. Deficiency Symptoms

Deficiency of calcium first of all brings about *bone abnormalities* like osteoporosis, osteomalacia and rickets as well as teeth abnormalities like dental decay, brittle teeth *etc.* Tetany (= nervous disorder reflected by intermittent tonic spasms) is also common. Tachycardia, excessive bleeding, hyperirritability, heart atony *etc.* are among other symptoms. Calcium ions are mobile and hence, they migrate from bones, teeth and other organs to those organs where deficiency is felt, resulting *deficiency* in the supplying organs.

6.2.6. Daily Requirement

Stated in Table 16 under the Chapter 9 (Dietetics).

6.2.7. Adverse Effect

High intake of calcium causes great harm to the body. Calcium ions in excess are toxic and causes renal calculi, renal insufficiency, cardiac arrythmia, atherosclerosis, low respiratory rate, severe weakness of muscles, stomach pain, anorexia, nausea, lethargy to induce coma, severe constipation *etc.* (Calcium ions, if directly instilled in the blood stream highly increases pumping action of heart and this should not be done as the patient may collapse).

6.3. Phosphorus

Phosphorus is the *second-most abundant mineral element in the body, next to calcium.* This non-metallic element is found in every cell in the body and it exists in combination with alkalis. It remains as a *buffer in cellular fluid*, by maintaining proper balance of acidity and alkalinity. Along with calcium, it forms phospholipids, phosphoproteins, nucleic acids, enzymes and macroergic bonds (ATP). About 85 per cent of phosphorus in the body is present in bones and teeth along with calcium, 10 per cent in muscles and rest in soft tisses. As stated under the heading on calcium, calcium and phosphate ions are complementary. Product of phosphate and calcium ions usually remains constant. Calcium ulilization takes place along with phosphate ions.

6.3.1. Functions

Among many function of phosphorus in the body, some pertinent ons may be stated as follows:

(i) Phosphorus is an integral component of atp, dna, rna, nad, nadp, fad, glycerophosphate *etc.*
(ii) It is necessary for producing energy in the cell.
(iii) For development of bones and teeth, phosphorus is the essential element together with calcium.
(iv) The element is essential for structure of cell-membrane.
(v) Maintenance of acid-base balance in the cells is accomplished by the element.
(vi) It is needed for production of energy and storage.
(vii) Production of protein is facilitated by the element.
(viii) It activates enzymes and hormones.
(ix) Many of the B vitamins are active in presence of phosphate ions.
(x) Phospholipids help to transport other lipids in blood.
(xi) It helps to carry oxygen to tissues.
(xii) Along with calcium, phosphorus helps muscles to relax and contract.
(xiii) Functioning of the nervous systems is improved by phosphorus.

6.3.2. Principal Food Sources

Adequate amount of phosphorus is present in many foods. However, the element is very rich in egg-yolk, milk, sea-fishes, red meat, cheese *etc.* Among plant source foods, it is abundant in peas, bean, other pulses, fig, walnut, raisin, pumpkin, whole grains *etc.*

6.3.3. Chemical Aspects

Phosphorus which comes under Group V – A of the Periodic Table is a solid matter and non-metallic element with atomic number 15, atomic weight 30.9738 and valence as 3.5. It has several allotropic forms, designated by colour.

6.3.4. Absorption

Absorption of phosphates takes place mostly from duodenum and some from other parts of small intestine. Vitamin-D is *essential for absorption* of phosphate ions.

6.3.5. Deficiency Symptoms

Phosphate is much present in plenty of foods and hence, deficiency of phosphorus is rarely found in the body. Secondary deficiency is sometimes observed (hypophosphataemia) due to intake of antacids. However, some deficiency symptoms of phosphate are, retarded growth, loss of weight, imperfect bones and teeth, continuous fatigue, perverted appetite *etc.* but deficiency symptoms are rarely observed. At old age, deficiency of phosphorus and calcium is sometimes observed in some people because of improper diet.

6.3.6. Daily Requirements

Daily requirement of phosphorus for infants is 400 mg, for children 900– 200 mg, for adolescents 1200 mg and in adults, it may be 1500 mg. Requirement may be 1500–1600 mg for pregnant women and upto 1800 mg for the lactating mothers. Normal level of phosphorus in blood plasma is 3–5 mg per decilitre. Urinary excretion of phosphorus takes place to about 1–2 gram in 24 hours.

6.3.7. Adverse Effects

Phosphorus has very low toxicity and undesirable symptoms due to excess intake of the element is hardly noticeable. Renal impairment may sometimes result due to excess phosphate intake, particularly in advanced age and this may increase the risk of demineralization. Excess calcium intake brings about magnesium deficiency. Calcium ions are in fact, antagonistic to those of potassium, sodium and magnesium.

6.4. Iron

Iron has the greatest role in the body in *synthesis of the compound, haemoglobin.* This is an iron-containing pigment which is present inside the red blood cells of blood. The function of the pigment is to *pick up oxygen from the lungs and supply that to the cells in all parts of body, i.e., to make them oxygenated.* (It should be remembered that oxygenation means making provision of oxygen and not making chemically oxidized). Without supply of oxygen to the cells, they cannot survive any more. An adult person contains iron to about 4 gram (little less in a woman) and of this, about 70 per cent is present in haemoglobin. One gram of haemoglobin takes up 1.34 ml of oxygen from the lungs.

Haemoglobin actually contains two part. These are called, *haem* and *globin.* Hcam is the *non-protein part which contains iron* and globin is the protein part (= histone protein) and it does not contain any iron. In a haemoglobin molecule, haem is present to 4 per cent and the globin part, *i.e.*, histone protein (which is actually a basic protein) constitutes 96 per cent.

Not only in the haemoglobin but iron is a constituent of many other compounds as well and for activity of enzymes in the body. These are cited below:

It is present in *myoglobin*, which is an iron-containing protein present in the muscles. In some *enzymes* (= haem-enzymes), *e.g.*, mitochondrial cytochromes and microsomal cytochromes (*cyto* - cell, *chrome*-colour), which have importance in cell respiration (= cytochrome transport system), iron is present. Besides these, in other enzymes also, *viz.*, catalase, peroxidase and the flavin-enzymes, *e.g.*, succinic dehydrogenase, xanthine oxidase, aconitase, DPNH-cytochrome C reductase, indophenol oxidase *etc.* remains to be a component.

Iron has its role in the transport and storage of some compounds and thus, it contitutes compounds like transferrin, ferritin, hemosiderin *etc.*

6.4.1. Functions

It has been stated under the above heading that most important function of iron is to synthesize haemoglobin and thus, supplying oxygen to the cells in the body. However, iron acts in not only synthesizing haemoglobin but is necessary for *formation and maturation of red blood cells* also. Haemoglobin is in fact, a large carrier of oxygen as one gram of it can pick up 1.34 ml of oxygen and delivers that oxygen to the cells. The iron-containing myoglobin protein which takes up oxygen acts in the muscles as an oxygen storage device. The iron-containing cytochrome has essentiality for oxidation of cells. The chromatin in the nucleus of cells contains iron which has the role in functioning of the nuclei. Above all, iron has a great role in the immune system in the body.

6.4.2. Principal Food Sources

Among animal source foods, meat, egg-yolk and milk, especially cow's milk are rich sources. Liver in meat is a highly rich source of iron. (Many quacks having not adequate background of medicine or nutrition sometimes. advise to take raw, *i.e.*, uncooked or unboiled liver of mutton making it mashed and recommend to the patients suffering from iron-deficiency anaemia. It should be noted that it is a highly dangerous practice. It is extremely difficult for human beings to digest even a very small piece of uncooked mutton liver, besides that, the liver may contain worms also. Hence, liver of mutton, chicken or other animals should be taken after thorough cooking only). Among plant source foods, bajra, ragi, nuts, date, fig, whole grains, dark green leafy vegetables contain high amount of iron.

6.4.3. Chemical Aspects

Iron is a solid from of metallic mineral element that comes under the Group VIII of Periodic Table. Its atomic number is 26, atomic weight is 55.847 and valence is 2, 3.

6.4.4. Absorption

Iron is absorped almost all through the gut but largely from the upper part of small intestine and *particularly in duodenum*. On absorption from food, the irons are passed into the blood and through the blood stream, they reach the bone-marrow where they are utilized in synthesis of haemoglobin.

Iron which is present in the food and particularly in the organic form of the food is mostly in the ferric (Fe^{+++}) form. But ferric form of iron is not absorbed and for absorption, it needs to be converted to the ferrous (Fe^{++}) form. Some substances in the food such as vitamin C, glutathione, amino acids having SH groups, chlorophyll *etc.* greatly help in conversion, *i.e.*, reduction of iorn from ferric to ferrous form. However, after the iron is absorbed in the body, it is again converted to the ferric form.

Absorption, *i.e.*, bioavailability of iron in the body also depends on a number of factors. The most important factors are requirement of iron in the body, growth, menstrual flow, pregnancy, lactation, blood loss, anaemia, ferrous or ferric form, solubility *etc.*

6.4.5. Storage

After the iron is absorbed in the body in the ferrous form, it is quickly converted into the *ferric form*. The ferric form of iron reacts with phosphate to form ferric-hydroxide phosphate. This compound combines with a protein known as apoferritin. As a result, an iron-phosphorus-protein compound is formed. This is called ferritin. *Iron is stored in the body mainly in this compound, that is, ferritin.*

However, apart from ferritin, iron is also stored in the body in the form of another compound, *which is known* as *haemosiderin*. But much less amount of iron is stored in this compound as compared to ferritin. This compound is not soluble in water unlike ferritin which is water-soluble. Storage of iron as ferritin takes place mostly in liver but also in spleen and bone-marrow. Iron stored as hemosiderin is actually not available to the body. Hence, ferritin is the principal compound, *i.e.*, source of storage of iron in the body for the supply when the body needs.

6.4.6. Deficiency Symptoms

Many people do not apparently show symptoms of iron deficiency although in reality, many are deficient in iron. It is observed that iron deficiency is much more common in case of adolescent girls, women who have attained child-bearing age and people of vegetarian food habit.

Deficiency of iron results in *iron-deficiency anaemia which is characterized by microcytic hypochromic red blood cells*. More than 2 billion people in the world suffer from iron-deficiency anaemia. The symptoms in case of women having the child bearing age are fatigue, breathlessness in slight exertion, giddiness, pallor of the skin and swelling of feet and ankles due to oedema. In the young children and infants, growth is retarded, appetite is less and skin becomes pallor in appearance. (If a child below 4 years of age is found to eat mud or burnt clay, there is possibility to apprehend that he or she has been suffering from iron-deficiency anaemia and in order to get iron in the body, such a practice is done without knowledge of physiology).

Recycling of iron in the body is a fact that deserves interest. As stated earlier, an adult person contains about 4 gram of iron in the body and about 70 per cent of iron exists in haemoglobin. Thus, iron content as haemoglobin in an adult may be about 2.8 gram. Haemoglobin exists in the red blood cells (RBC). The RBCs are not permanent structures. An RBC lives upto about 120 days and after that, it *dies* (destructed). When an RBC dies, the iron that is contained in the haemoglobin part of it evidently becomes free. This free form of iron is stored (as ferritin) in the body and then again re-used to form new haemoglobins.

Hence, theoretically speaking, *iron that is contained in the blood is not lost but it is recycled*. Thus, it seems that supply of iron through food is not necessary. However, loss of RBC, *i.e.*, haemoglobin (contained in it) has also been there. This takes place through faeces but to a very less amount. In case of women however, loss of haemoglobin, *i.e.*, iron is considerably high through menstrual flow and in pregnancy and lactation. But a question arises, *i.e.*, how much of iron is regularly lost in a healthy man who has not been suffering from *e.g.*, tract ulcer, malignancy, piles, other blood loss disease or has not sustained blood losing injury or surgical operation, such that adequate dietary iron is essential for him to maintain health?

6.4.7. Daily Requirement

Stated in Table 17 under the Chapter 9 on Dietetics.

6.4.8. Adverse Effects

Excess accumulation and absorption of iron in the body produces adverse effects. A physiological disorder, *haemochromatosis* may result due to excess of iron, which is however, a genetically controlled abnormality. Symptoms of this disorder are, loss of weight, abdominal pain, arthralgias (joint pain), weakness, palpitation, male impotence *etc.* It is also associated with pigmentation of skin and diabetes. High intake of iron reduces absorption of zinc as well and thus, produces zinc deficiency symptoms.

When there is high accumulation of iron in the body as may be considered to be a disease, even removal of some blood from the patient's body may become necessary, which is known as phlebotomy and this should be continued till the haematocrit comes to normal level. A report (appeared in the Telegraph Daily, 18 September, 2019) published in the Journal, *Neurology* has brought forward that excess haemoglobin in blood has been found to cause dementia and Alzheimer's disease.

Excess accumulation of iron in the body is usually not noticeable at the present time in India as cooking utensils made of stainless steel and aluminium are much used than iron utensils and drinking water obtained from hand-pump is less used. It may be pointed out in this connection that injection of the solution of iron salt directly into the blood stream which was a practice in olden days to cure anaemia does not get much support now to avoid heavy metal injury.

6.5. Sodium

Healthy human beings barring neonates cannot think of taking a diet without addition of common salt. Common salt which is chemically, sodium chloride (NaCl) contains 40 per cent sodium and 60 per cent chlorine. Common salt not only adds taste to the food but provides sodium and chloride ions in the body which are of help to the body. Sodium is a systemic electrolyte in the body and is a nutrient. It constitutes 0.1 per cent of body weight and as sodium chloride, it remains as 550–650 mg per 100 ml in blood plasma.

6.5.1. Functions

Sodium is the *main electrolyte of extracellular fluid.* With potassium, it is essential in coregulating ATP. It is needed in the body for proper fluid balance, nerve transmission and muscle contraction. The sodium as the mineral has its necessity in maintaining osmotic pressure and acid-base balance. For absorption of water and other nutrients, necessity of sodium is felt. It serves the body both as sodium ions and chemical compounds. The mineral maintains normal rhythm of heart beat, along with calcium and potassium, conduction of nervous impulse, blood pressure and clotting of blood.

6.5.2. Principal Food Sources

Common salt (= table salt) which is chemically sodium chloride is the most important source to get sodium in the body. Milk, oysters, some fishes, sea-fishes, sea-foods, spinach *etc.* also contain appreciably high amount of sodium.

6.5.3. Chemical Aspects

Sodium is a soft and solid metallic element having a strong affinity for other non-metallic elements including oxygen. It comes under Group A – I of Periodic Table, having atomic number 11, atomic weight 22.98977 and +1 valence.

6.5.4. Absorption

Most of the sodium obtained through sodium chloride which is common salt and chemically NaCl is filtered and re-absorbed when sufficient aldosterone is present in the distal nephron, Sodium is re-absorbed except in thin segment.

6.5.5. Storage

Salts or ions of sodium are not stored long in the body. In relatively short period, they are excreted through urine, faeces and sweat.

6.5.6. Deficiency Symptoms

Deficiency symptoms of sodium are rarely met with as people take adequate common salt in their diet. Salt intake is relatively high in India and also in some other countries and may be in the U.K. Nevertheless, deficiency of sodium, known as hyponatraemia is sometimes met with when there is heavy loss of sodium from the body, *i.e.*, hypovolamic condition. Symptoms of deficiency are weakness of muscles, loss of body weight, impaired digestion, nerve disorder, paralysis, palpitation and fall of blood pressure. Indiscriminate use of anti-hypertensive and diuretic drugs may be a common cause of sodium deficit in the body.

6.5.7. Daily Requirement

Daily intake of common salt, *i.e.*, sodium chloride should not be less than 5–10 gram in an adult person and it should be less in children.

6.5.8. Adverse Effects

Excess intake of common salt causes hypernatraemia and is harmful to the body. Common symptoms are less consciousness, low urinary output, hyperaldosteronism, heart failure *etc.* But it causes great harm to the renal functioning and may be dangerous for people suffering from high blood pressure as it quickly raises the arterial blood pressure level. (It may be mentioned that use of common salt is strictly restricted among many ascetics of the Hindu religion as salt intake in the body even to a slightly higher level is believed to cause great harm in functioning of organs and worshipping).

6.6. Potassium

Potassium is a systemic elecrolyte and is the major intracellular cation. It constitutes 0.35 per cent of body weight. Potassium ions are of great value in cardiovascular functioning and salt of potassium are much advised in many ailments.

6.6.1. Functions

Potassium is essential in co-regulating ATP with sodium. It regulates water and fluid in the body and maintains electrolytic balance. Along with sodium and chloride, potassium ions maintains acid-base balance and osmotic pressure. For functioning of muscles and heart and maintaining cardiac pumping in rhythm, it is necessary. Potassium decreases blood pressure by enhancing loss of sodium through urine. For electrical impulses which are necessary in functioning of nerves, potassium ions are necessary.

6.6.2. Principal Food Sources

Fruits and nuts are rich sources of potassium. It is also abundant in potato, sweet potato, tomato, beans, lentils, milk, sea-fishes *etc.*

6.6.3. Chemical Aspects

Potassium is a solid form of mineral element coming under Group I – A in the Periodic Table. Atomic number of potassium is 19, atomic weight is 30.098 and of 1 balance.

6.6.4. Absorption

Potassium ions are not much lost and are excreted at about 2 - 4 gram per day through urine. They are excreted from the distal nephrons. Much of filtered potassium ions are however, re-absorbed in the proximal part of convoluted tubules.

6.6.5. Deficiency Symptoms

Deficiency of potassium in the body is known as *hypokalaemia*. This usually takes place in severe diarrhoea and some enteric diseases. The common symptoms are general weakness of the body, giddiness, extreme thirst, mental confusion, fall in blood pressure, low output of urine *etc.* In extreme form of hypokalaemia, there may be acute renal failure and also heart failure. Bone may lose strength. Sterility in men and women may also result in acute deficiency of potassium for long period.

6.6.6. Daily Requirement

Daily requirement of potassium for adults may be 4–6 gram. High potassium intake is advised for cardiac patients but not too much.

6.6.7. Adverse Effects

Due to excess intake of potassium which usually results when excess potassium intake in advised in diarrhoea, metabolic acidosis or when diuretic drugs are administered, kidney failure may result. Some other symptoms of potassium excess are general weakness, paralysis and impairment in electrical conduction in the heart, ventricular fibrillation and may even lead to death. Intravenous supply of potassium and also calcium ions may be extremely harmful and is not advisable unless there is genuine need and should be given only under expert medical supervision.

6.7. Magnesium

Magnesium is a very important mineral element in *bone formation* and strength. About 70 per cent of magnesium is present in bones. Magnesium ions are *interrelated to those of calcium, potassium and sodium* in many physiological reactions. They are sensitive to stress.

6.7.1. Functions

Essentiality of magnesium in bone formation is well established. It is also of great help to the teeth as the magnesium ions promote resistance to decay of teeth by holding them in the teeth. Magnesium activates plentiful enzymes in the body as metallic co-factor including enzymes concerned with DNA replication and RNA synthesis. It plays part in glycolysis and formation of cyclic AMP. In secretion of parathyroid hormone, magnesium has its necessity. Magnesium boosts up immune system and is necessary for nerve and muscle functioning. It reduces blood sugar in type-2 diabetes and blood pressure in hypertension. A report states that on interacting with neuro-transmitter, magnesium favours sleep.

6.7.2. Principal Food Sources

Magnesium is much in existence in many food-stuffs. Green parts of plants always contain magnesium and hence, spinach and other green leafy vegetables are rich sources. Nuts, legumes, many fruits, oats, milk, butter, red meat, fish *etc.* also contain abundant magnesium.

6.7.3. Chemical Aspects

Magnesium is the solid form of element under the Group II – A of Periodic Table. Atomic number of the element is 12, atomic weight 24.312 and is of 2 valence.

6.7.4. Absorption

Magnesium is absorbed from the upper part of intestine. The element is absorbed with great difficulty as it forms phosphate compound like that of calcium.

6.7.5. Deficiency Symptoms

Deficiency of magnesium, *i.e., hypomagnesaemia* is rarely noticed in people as most foods contain high content of magnesium. Nevertheless, symptoms like diarrhoea, vomiting, neuromuscular disturbances, tetany *etc.* are sometimes noticed in deficiency. Symptoms are more common among alcoholics.

6.7.6. Daily Requirement

Staple diet provides sufficient magnesium in the body. However, daily intake of 300–400 mg of magnesium is advised if the diet is not proper. Amount necessary for adult women may be less than men.

6.7.7. Adverse Effects

Hypermagnesaemia due to excess intake of magnesium may take place but this is a rare occurrence. By excess intake, too much defecation (particularly by consuming magnesium sulphate) may be a common symptom. Muscular weakness, low blood pressure, bradycardia, hyperactivity, convulsion *etc.* are some of the effects of excess intake of magnesium. Very high intake of magnesium salts cause great harm and may even bring about cardiac arrest.

6.8. Sulphur

Sulphur is one of the important macro-mineral elements in the body. It is present in blood to about 5 mg in 100 ml. The compounds of the element protect many of the organs in the body. Excretion of sulphur takes place to about one gram from the body daily through faeces, saliva and urine.

6.8.1. Functions

Sulphur is an essential component of three amino acids in the body, *viz., methionine, cystine* and *cystein*, which are ingredients of many important proteins. It is a co-factor of some enzymes and helps in functioning of co-enzyme A. It is necessary for sulpholipid formation in nerve tissue and also in synthesis of B-vitamins, *viz.*, thiamine and biotin. The radicals of sulphur, *i.e.*, sulphydryl radicals (SH) give rise to many important chemical compounds. Sulphur has necessity for functioning of sulphur containing protein, like mucin and cartilagenous proteins. The sulphur containing tripeptide, glutathione plays part in tissue oxidation.

Detoxification is stated to be an important role of sulphur. Cystine, the sulphur containing amino acid detoxifies bromobenzene and the sulphuric acid detoxifies indoxyl. Sulphur protects protoplasm of cells also. It is contained in heparin which protects vascular blood from clotting. Sulphur also gives rise to some hardness in elastic tissue in the body like nail, hair, cartilage *etc.* The compounds of sulphur are known to promote immune system.

6.8.2. Principal Food Sources

Much of the sulphur is taken by the body from foods of various types. It is abundant in many animal source foods like meat, poultry, egg, fish, milk and dairy products. It is also fairly high in the nuts.

6.8.3. Chemical Aspects

Sulphur is a yellow solid matter that actually occurs in elemental form. It comes under Group VI – A in the Periodic Table and has atomic number 16, atomic weight 32.06 with 2, 4, 6 valencies.

6.8.4. Deficiency Symptoms

Deficiency symptoms of sulphur are rarely observed. In acute deficiency, growth may be retarded due to inadequate production of sulphur-containing amino acids. Dermatitis and improper development of hair and nails are sometimes noticeable in acute shortage of sulphur in the body.

6.8.5. Daily Requirement

Daily requirement of sulphur in the body is not correctly assessed and the body gets the necessary amount from the ordinary mixed diet.

6.8.6. Adverse Effects

Sulphur is not consumed in excess and hence, adverse effects are not observed. Administration of sulphur drugs used for therapeutic purposes may sometimes produce abnormal symptoms in some people though not in all. Sensitivity to sulphur may be stated to be a cause to those people rather than excess intake of sulphur.

6.9. Chlorine

Chlorine is one of the nutritional elements in the body, although it is a *highly irritating and very poisonous gas and hence, cannot exist in the free form, i.e., as chlorine gas.* Accordingly it exists in the body as a binary compound of chlorine. In other words, it exists as a *salt* of chlorine, *i.e.*, as chloride. Existence in the form of hydrochloric acid, which is chloride of hydrogen (HCl) is the common form of its existence in the body and functioning.

Chlorides (rather than chlorine) have great role in the body. Juice of the stomach (= gastric juice) is formed from chloride. Gastric juice is a thin, colourless liquid which contains hydrochloric acid (HCl), pepsin, mucin, small amount of inorganic salt *etc.* It is strongly acidic in reaction (pH = 0.9 – 1.5) which is due to the presence of hydrochloric acid. Hydrochloric acid is aqueous solution of hydrogen chloride (HCl), containing 35–40 per cent HCl on weight basis.

Apart from stomach, the chlorides exist in the body also in the inside of the cells and as extracellular fluid. Chlorides provide two-third of the anions of blood plasma and is the *chief anion in the body.* Blood serum contains 350–390 mg of chloride per 100 ml.

6.9.1. Functions

As stated above, chlorine, in other words, chloride is a constituent of acid in the stomach in the form of hydrochloride (= hydrochloric acid). This acid cannot form unless sufficient chloride ions enter the body through salt or food. As stated under the heading of carbohydrate (Chapter – 2), digestion of carbohydrate and chemical breakdown (= acid-hydrolysis) of sugar will not take place unless adequate amount of hydrochloric acid is present in the stomach.

Apart from the above roles, chloride ions are of great necessity in maintaining acid-base balance in the body, regulating osmotic pressure and proper fluid balance. They are also essential for secretion of ptyalin, activity of pancreatic amylase and other enzymes, regulating and stimulating muscular activity, transmitting nerve impulse and have great role in many other physiological processes. Chlorides are strong electron acceptors and thus, are powerful oxidizing agents.

6.9.2. Principal Food Sources

Table salt (= common salt) which is chemically sodium chloride (NaCl) is by far, the only or the main source of getting chloride ions in the body. However, it is also present in many processed food, milk, leafy vegetables, bread *etc.*

6.9.3. Chemical Aspects

Chlorine, which is a gaseous form of the chemical element comes under the Group VII – A in the Periodic Table. Its atomic number is 17, atomic weight is 35, 453 and exists in the valencies of –1, +1, +3, +4, +5, and +7. The diatomic gas of chlorine, *i.e.*, its molecule (Cl_2) has molecular weight of 70.906.

6.9.4. Absorption

Chloride ions are readily absorbed from all parts of the small intestine. They are also re-absorbed in the body through renal tubules.

6.9.5. Storage

Chlorides are much retained in the body and loss of them through excretion is not high. More than one gram of chloride is usually not excreted from the body in a day. Retention of chlorides is much higher in the skin and in the sub-cutaneous tissue.

6.9.6. Deficiency Symptoms

Deficiency of chlorides is rarely noticeable in the body. However, when salt intake in the diet is severely restricted particularly in high blood pressure patients, deficiency may occur which is known as *hypochloraemia*, *i.e.*, very low level of chlorides (as also sodium) in the blood. It results in fatigue, lassitude, inactivity, impaired nerve functioning *etc.* and may be fatal if the person attempts to work hard in chloride deficient condition.

6.9.7. Daily Requirement

Daily requirement may be 10–12 gram table salt (= sodium chloride) that contains chloride.

6.9.8. Adverse Effects

Adverse effect due to excess intake of chloride is not clearly known but it can cause hyperchloremic condition. Long term use of medicines containing chloride may bring about such abnormality.

6.10. Iodine

Iodine is non-metallic element which is used in the body in traces. However, this trace element has a tremendous role in the body in being a constituent of the hormones that are secreted by the highly useful endocrine gland, which is *thyroid gland*. The hormones are, tri-iodothyronine (known as T3) and tetra-iodothyronine (known as T4). These are secreted by the follicular cells and the thyro-calcitonin, which is secreted by the para-follicular cells. The thyroid hormones are produced by the combination of iodine which is taken up from the plasma of blood and the residues of the amino acid, *tyrosine*. About 50 mg of iodine is present in a human adult and of this, about 15 mg is present in the thyroid gland.

Apart from thyroid gland, iodine is also taken up and used by some other organs, notably mammary gland, salivary gland, gastric mucosa and possibly by some other parts.

6.10.1. Functions

The thyroid hormones are produced in the thyroid gland. This endocrine gland is situated in front of the neck. This gland has two lobes which are connected by an isthmus. The thyroid hormones are such that low (= sub-optimal) production of them is harmful to health and causes the abnormality known as *hypothyroidism* while high (super-optimal) production causes the abnormality known as *hyperthyroidism*. These have been discussed later.

Some of the important functions of thyroid hormones are stated below in brief.

(i) These hormones play vital role to bring about *general growth* of the body. This is accomplished by synthesis of somatomedins, increased synthesis of protein and activation of growth hormone. Experiments conducted in lower forms of animals confirmed that growth of the body becomes severely retarded if production of thyroid hormones becomes very low.

(ii) Thyroid hormones are necessary for protein synthesis but if the hormones accumulate to a very high level due to internal production or due to supply from outside (which may be by taking high dose of thyroid medicine or high intake of iodine in diet), proteins that are already synthesized in the body may catabolise (= breakdown).

(iii) High secretion of thyroid hormones *increase basal metabolic rate* (BMR). This causes rise of body temperature but not fever.

(iv) Thyroid hormones have great role in *nervous system*. They are essential for development, maturation and functioning of the *neurons* (= nerve cells) and myelogenesis (= development of myelin sheaths during development of the central nervous system, producing in the marrow).

(v) Cardiac activity is increased due to increased response of catecholamines (= one of the two active amines, epinephrine and norepinephrine, derived from the amino acid, tyrosine. They have great role on the nervous and cardiovascular systems, metabolism, temperature regulation and smooth muscles).

(vi) Thyroid hormones cause high accumulation of sugar in the body plasma by higher absorption of sugar.

(vii) For growth and maintenance of skeletal muscles, thyroid hormones are of great help.

(viii) They have role in causing synthesis of lipids and their catabolism. By catabolysing (= breaking down) lipids, they tend to lower the cholesterol and triglyceride level in blood plasma.

(ix) In absorption of vitamin B-12 from the intestine, thyroid hormones play part.

(x) For conversion of beta-carotene to form vitamin A, thyroid hormones are necessary.

(xi) By preventing loss of moisture form skin, thyroid hormones bring lustre of the skin.

(xii) They are helpful in formation of red blood cells (= erythropoiesis).

(xiii) Stimulating effect of thyroid hormones in production of breast milk is known.

(xiv) An important function of thyroid hormones is to stimulate activity of other hormones, an effect which is referred to as *permissive action* of a hormone.

6.10.2. Principal Food Sources

Sea water is very rich in iodine. Hence, sea fishes, shell fishes, sea weeds are rich sources. Milk is a very good source. It is found that when cereals or other agricultural crops grow in such soil which contain high iodine, they take up much iodine from the soil and consumption of such foods provide iodine in the body.

6.10.3. Chemical Aspects

Iodine which is one of the less-active halogens is a black solid matter and is classed as a non-metal. It comes under the Group VII – A of the Periodic Table, has atomic number 53, atomic weight 126.9045 and of 1, 3, 5 and 7 valencies.

6.10.4. Absorption

Iodine is readily absorbed from the small intestine. Absorption of the nutrient is arrested by the nitrate ions. Some vegetables like cauliflower, cabbage also interfere utilization of iodine.

6.10.5. Deficiency Symptoms

Thyroid hormones that are secreted by the thyroid glands are such, that if their production is low, *i.e.*, sub-optimal, it causes disease and on the other hand, if the production of thyroid hormones is high, *i.e.*, super-optimal, then also it causes disease. The former case, *i.e.*, when the production of thyroid hormones is less, the abnormality is called *hypothyroidsm* and when the production is more than what is required, the abnormality is termed as *hyperthyroidism*. It should be noted that both are very serious health abnormalities. Discussion of the thyroid glands, thyroid hormones, hypothyroidism and hyperthyroidism have been nicely made by A. B. Singha Mahapatra is his book, Essential of Medical Physiology (2005), (Published by Central Book International, Kolkata and Mumbai) and the following presentation is made, largely based upon discussion made in the said title.

Hypothyroidism

This refers to sub-optimal production of thyroid hormones, which is the result of deficiency of iodine in the diet, or due to inadequate absorption or ultilization of iodine in the body even though it has been taken up in adequacy.

It is estimated that 120–150 million people around the world suffer from hypothyroidism induced by iodine deficiency. About 12 per cent of people in the world are known to suffer from thyroid gland malfunctioning at least at some part in their lives. Women are considered to suffer from thyroid hormone deficiency about eight times more than men.

Goitre is a common symptom of both hypothyroidism and hyperthyroidism. This abnormality is a condition that results due to improper functioning of the thyroid gland. Hence, an understanding of it becomes necessary before going to further discussion. The term, goitre (synonym : *struma*) is actually derived from the Latin word, guttur which means an *enlargement of the thyroid gland.* That is, when there is a swelling of the thyroid gland and if the swelling is not due to any inflammation or neoplasm, it is referred to as a goitre. Goitres may be of several types as stated below: (i) Simple goitres (= euthyroid type), which may be collodial when colloid level becomes high, parenchymatous when there is increase of follicles and nodular goitre when some nodules of the gland are only swollen. (ii) Hypothyroid goitres when the thyroid gland becomes large but secretion of thyroid hormone is less, *e.g.*, cretinism and myxodema. (iii) Hyperthyroid goitres when the thyroid gland becomes large but secretion of thyroid hormone is also very high, *e.g.*, Grave's disease and nodular goitres and (iv) Physiological goitre, when thyroid enlargement occurs in girls at their puberty.

In case of hypothyroidism, goitre occurs due to insufficient production of thyroid hormone, which is due to insufficient intake of iodine or its inadequate absorption in the body.

Hypothyroidism may be of primary or secondary type. In primary type, either adequate hormones are not produced or thyroid hormones are adequately produced but destructed. Secondary hypothyroidism occurs due to improper functioning of the pituitary or hypothalamus gland.

Abnormalities that take place due to hypothyroidism may be of two types. These are called *cretinism* and *myxedema*. In cretinism, deficiency of thyroid hormones takes place in a new-born baby since birth. Symptoms are not visible as long as the infant sucks mother's milk because, mother's milk provides the necessary requirement of thyroid hormones. But when the infant stops sucking of mother's milk, abnormal symptoms of cretinism appear in him or her. These are, retarded growth, idiotic appearance of face, too much saliva dribbling, retarded mental and sexual development, less appetite, less resistance to disease infections, less basal metabolic rate, dryness of skin, yellow patches on skin due to deposition of carotene pigment and so on.

Myxodema occurs in adult persons. Symptoms are, swelling here and there on face due to oedema. Swelling may be on other parts of the body also in severe deficiency of the thyroid hormones. Voice and heart rate are slow. Mental ability is impaired, blood cholesterol is increased and besides these, there are also many other symptoms of myxodema that may appear.

It is sometimes stated that hypothyroidism is less prevalent among people living around sea-coasts as they take plenty of sea-fishes which contain high iodine while the incidence is very high is people inhabiting in mountainous area and the upcountry where availability of dietary iodine is much less. However, there is no data-based evidence of this concept. At the present time, Govt. of India has made it mandatory to take iodized salt to the people of all parts of the country, regardless of their place of living. It may be noted that even if, intake of iodine becomes somewhat higher, it will have no adverse effect as excess iodine is not absorbed in the body.

6.10.6. Daily Requirement

Daily requirement of iodine which is necessary for a person has not been assessed with exactness but 100–200 microgram per day are recommended and the amount should be higher in children and particularly in pregnant women.

6.10.7. Adverse Effects

Excessive production of thyroid hormones may cause abnormality which is known as *hyperthyroidism* or *thyrotoxicosis*. This may be due to too much intake of iodine or increased functioning of the pituitary or hypothalamus gland. Excess intake of iodine is usually not the cause as high dietary iodine is not absorbed from the small intestine. However, possible causes are, enlargement of the thyroid gland which secretes very high amount of thyroid hormones, very high secretion of the thyroid stimulating hormone (TSH) from the pituitary gland which causes to increase production of thyroid hormone or secretion of thyroid stimulating immunoglobulin (TSI).

Adverse effects of hyperthyroidism are many. Grave's disease is an abnormality when goitre, *i.e.*, swelling of thyroid gland and protrusion of the eye balls, which is known os *exophthalmos* appear together. Protrusion of the eye balls becomes such that whole of the cornea would become visible. Other abnormalities of hyperthyroidism are, high blood sugar, cardiac output, increased basal metabolic rate, appetite, irritability, loss of weight, high pulse rate during sleeping *etc.* Nervousness, hyperactivity, tremor, high temperature with fever, osteoporosis, fatigue, diarrhoea are some of the other symptoms.

6.11. Zinc

Zinc acts as a metallic co-factor in hundreds of enzymes in the body. It is a *potent booster for growth*. It plays part in regulation of blood glucose by interacting with insulin hormone. Zinc is distributed in all parts in the body. The nutritive element is also credited to boost up immunity in the body.

6.11.1. Functions

Zinc acts as a metallic co-factor in plentiful enzyme-systems. Some of the notable enzyme-systems are caroboxypeptidase, liver alcohol dehydrogenase, carbonic anhydrase *etc.* It helps metabolism of carbohydrate, protein and fat. On interaction with the hormone insulin, zinc regulates sugar levels in blood. It helps to release pancreatic enzyme in digestive function. Zinc has a great role in cell division and thus, is of great help in repair of wounds. For structure and function of skin, zinc is an important element. It occurs with proteins in many organs. For releasing vitamin A which is stored in the liver, zinc plays part. For sperm production and healthy pregnancy, zinc has its great importance. The element is of help in maintaining vision and smelling power of a person. It has a role in proper development of skin and hair. In boosting up immunity in the body, the element has a great role.

6.11.2. Principal Food Sources

Zinc is highly abundant in red meat, oysters, poultry, eggs, cheese, shell fish, fishes, milk, whole grains, legumes and many fruits.

6.11.3. Chemical Aspects

Zinc is a white, crystalline metal belonging to the Group II – B of the Periodic Table. Its atomic number is 30, atomic weight is 65.38 and has 2 valencies.

6.11.4. Absorption

Absorption of zinc takes place in small intestine. Nitrates, phytates and oxalates in foods reduce absorption and availability of zinc in the body. Sometimes, hand-made breads (= ruti, chapati) bind and reduce absorption of zinc. Oxalate present in spinach and rhuburb also bind zinc and makes it unavailable to the body.

6.11.5. Deficiency Symptoms

About 2 million people in the world suffer from zinc deficiency. Although amount of zinc in food may be high but its availability in the body is a problem in most cases.

In children, growth is stunted and skin infections are very common in deficiency of zinc. In adolescents, mental and sexual development are impaired and there in loss of appetite. Puberty is delayed. Skin infections and alopacia are very common. Deficiency may lead to some cancers, like that of prostrate, breast and skin. Hydrochloric acid content in stomach may be decreased in zinc-deficient condition.

6.11.6. Daily Requirement

Daily requirement of zinc is adult persons should be 10–15 mg. For infants, it should be 5–6 mg and it should be 15–20 mg during first six months of pregnancy.

6.11.7. Adverse Effects

Excess of zinc in the body is toxic. It interfers copper metabolism. Toxic effects of zinc salt intake are metallic taste, burning of the mouth, burning of throat and oesophagus. Blood comes out with vomiting. Washing of stomach should be done as first-aid treatment.

6.12. Copper

Copper is considered as the third most abundant trace metal after iron and zinc. Human body contains a total of 100–150 mg of copper. The mineral has a great role in *haemoglobin synthesis*, absorption and utilization of iron and activity of many redox enzymes. For boosting immunity, brain functions and other health giving properties, the mineral is useful.

6.12.1. Functions

Copper has a number of physiological functions. It is a metallic co-factor for many *redox enzymes* including cytochrome-c oxiase, tryptophan dioxygenase, cytochrome-c superoxide dimutase, ferroxidase *etc*. The trace element is essential for utilization of iron efficiently. In the synthesis of haemoglobin, it is necessary for the formation of metalloporphyrin. Copper has great role in functioning of brain. In processing of the collagen protein, it is of great help and hence, it is of value in skin, bones and other connective tissues. For proper functioning of circulatory system and preventing heart disease, the mineral has importance. It helps in lipid metabolism. The mineral boosts immunity in the body as well.

6.12.2. Principal Food Sources

Liver, kidney, sea fishes, oyser, nuts and whole grains are abundant sources of copper.

6.12.3. Chemical Aspects

Copper is a reddish metal under the Group I–B of Periodic Table and has atomic number 29, atomic weight 63.546 with valencies of 1 and 2.

6.12.4. Absorption

Absorption of copper is very less and it takes place from the small intestine.

6.12.5. Deficiency Symptoms

Symptoms of copper deficiency are extremely rare. In deficiency, immune system is weekend and *anaemia* (normochromic, microcytic) appears, which is due to improper absorption of iron. General weakness, impairment of central nervous system, impaired respiration, osteoporosis *etc.* are other symptoms. In some people, a genetic disease, known as Menke's disease may result due to inability to absorb copper. This leads to severe impairment of mental development.

6.12.6. Daily Requirement

Daily requirement is not correctly assessed but recommended dose is 2 mg which is obtained from diet.

6.12.7. Adverse Effects

Although excess copper in the body is toxic but adverse effect due to high intake of copper is rarely met with.

In some people however, a *hereditary syndrome* appears in which, copper accumulates in relatively high amount in various organs and notably in brain, liver, kidney and cornea. The abnormality is known as Wilson's disease (Syn. Hepatolenticular degeneration, or Westphal-Strupell pseudosclerosis) and it is due to *high intestinal absorption of copper*. The abnormality is characterized by degenerative changes in the brain, cirrhosis of the liver, tremor splenomegaly, mental confusion, emaciation, involuntary movements, dysphagia and so on. As there is no effective treatment, copper intake to the patient should be severly restricted. Copper binder, *e.g.*, D-penicillamine may be administered orally aiming at making reduction of copper in the various tissue (It should however, be noted that the copper binder, D-penicillamine may bring about deficiency of iron and pyridoxine).

6.13. Cobalt

Cobalt has its great significance in that, vitamin B-12, *i.e.*, cyanocobalamin (= cobamide) contains in its molecular cobalt cyanide and amino group. Thus, cobalt is a constituent of vitamin B-12. It remains as 4–5 per cent of vitamin B-12.

Cyanocobalamin exists in three forms. In one form, cobalt remains bound to cyanide and in the other two forms, it remains bound to the hydroxyl and nitrite groups respectively. Nitrocobalamin group has also been there in which nitrogen group is present in place of cyanocobalamin.

6.13.1. Functions

As stated above, cobalt is a constituent of vitamin B-12. It is essential for homopoiesis. Cobalt stimulates production of red blood cells, but for this effect, its use as a therapeutic agent is not advised in anaemia.

6.13.2. Principal Food Source

Normal diet contains adequate cobalt which is necessary to provide the requirement of the human body.

6.13.3. Chemical Aspects

Cobalt, the grey, hard ductile metal is alike to iron and nickel in free- and combined-forms. It comes under the Group VIII of the Periodic Table. Its atomic number is 27, atomic weight 58.9332 and is of +2 and +3 valencies.

6.13.4. Deficiency Symptoms

Deficiency of cobalt causes anaemia in ruminants of animals but not demonstrated in human body.

6.13.5. Daily Requirement

Daily requirement of cobalt is not worked out but necessary requirement is obtained by a person through diet.

6.13.6. Adverse Effects

Accumulation of excess cobalt in the body is harmful. It may cause degeneration of the *alpha cells* of Langerhans. Excess accumulation may also cause anorexia, nausea, vomiting, difficulty in hearing, hyperplasia, compression of trachea *etc.* By making a stimulation of erythropoietin hormone which steps up erythropoiesis, excess cobalt in the body may increase number of red blood cells (= cobalt-induced polythemia).

6.14. Manganese

Manganese acts as an important metallic co-factor in many enzyme-systems. It is also needed for normal bone development. The mineral is essential for overall management of body.

6.14.1. Functions

For action of many enzyme systems, manganese is essential as a co-factor. It is associated with protein and fat metabolism. For bone development in children and to prevent osteoporosis in adults, this trace element is necessary. Proper functioning of pituitary gland, kidney, pancreas and liver is accomplished by manganese. The element promotes healthy nervous system and digestion. It boosts up immunity system as well.

6.14.2. Principal Food Sources

Manganese is present is many food-stuffs. Whole grains, legumes, banana, other fruits and vegetables and also tea and coffee are important sources of manganese.

6.14.3. Chemical Aspects

Manganese is a hard, solid metal of Group VII – B of the periodic table. Its atomic number is 25, atomic weight 54.938 and is of 2, 3, 4, 6, 7 valencies.

6.14.4. Absorption

Absorption of manganese from different foods is estimated to be 5–10 per cent. It has been commented upon by the Indian Council of Medical Research – National Institute of Nutrition, Hyderabad, "Surprisingly, diets of poor income groups have a better absorption of manganese than diets of higher income groups. The basis for these differences in absorption is not known."

6.14.5. Deficiency Symptoms

Manganese deficiency symptoms are not much common. In children, retarded growth and improper bone development may be noticed, if deficiency is severe. Fatigue, impaired functioning of brain and reproductive system may take place due to manganese deficiency. Deficiency may also result in hypochromic anaemia.

6.14.6. Daily Requirement

Not exactly known but 2–3 mg seems to be optimum which is available from normal diet.

6.14.7. Adverse Effects

Any serious problem due to excess intake of manganese is not met with. Sometimes abnormal symptoms are noticed among people working in manganese mines. The symptoms are muscular weakness, tremor, central nervous system disorder *etc.*

6.15. Selenium

Selenium has great role for its importance as an *anti-oxidant* rather than as a trace-element. However, the element has also many health benefits and therapeutic effect against some ailments.

6.15.1. Functions

Most important function of selenium is, it is a component of some of the anti-oxidative enzymes, particularly glutathione peroxidase and thereby, it protects the body against oxidative damage, *i.e.*, it serves as a *potent anti-oxidant.* Selenium tends to utilize iodine in production of thyroid hormone. It helps in functioning of heart. The element has a role in preventing some types of cancer. Boosting up immunity of the body is also accomplished by the element.

6.15.2. Principal Food Sources

Some foods like Brazil-nut, sea fishes, meat, egg, grains *etc.* are good sources of selenium. Cereals and millets when grown in selenium-rich soil take up much selenium from the soil and thereby, their content of selenium is increased.

6.15.3. Chemical Aspects

Selenium is a solid metal resembling sulphur and comes under the Group VI–A of the Periodic Table. Its atomic number is 34, atomic weight is 78.96 and has valencies of 2, 4 and 6.

6.15.4. Deficiency Symptoms

An endemic heart disease known as *Keshan disease* occurs in children and women of child-bearing age in some parts of China in particular, by consuming cereals which are grown in selenium-deficient soils. Selenium deficiency may result in necrosis of liver and some types of cancer also.

6.15.5. Daily Requirement

Not correctly assessed but 0.055 mg seems to be satisfactory and this is obtained from normal diet.

6.15.6. Adverse Effects

Too much accumulation of selenium in the body, known as *selenosis* is toxic. Some effects of selenosis are brittle nails and hairs, skin lesions, nausea, vomiting, abdominal pain, muscular spasm, restlessness *etc.* and a garlic like odour comes out from mouth.

6.16. Fluorine

Fluorine, in other words, fluoride is given importance for its effect on *protection of teeth* and has not much importance as a nutrient.

6.16.1. Functions

Fluoride is known to have a great role in teeth and bones. It protects teeth from decay and dental caries and also cures dental abnormalities. Addition of fluoride in water has nowadays become a practice in some countries. This fluoridification of water is done to protect teeth and bones but is not always recommended as it leads to many hazards.

6.16.2. Principal Food Sources

Under the condition of India, water contains adequate fluoride.

6.16.3. Chemical Aspects

Fluorine, which is the gaseous chemical element is most reactive of all the halogens. It belongs to the Group VII – A of the Periodic Table and has atomic number 9, atomic weight 18.9984 and has - 1 valence. Body gets fluorine in the form of fluoride, which is a salt of hydrofluoric acid.

6.16.4. Deficiency Symptoms

Deficiency symptoms of fluoride is not observed under Indian condition. However, topical application of fluoride on gums is of benefit to the teeth. This is applied as stannous fluoride, sodium fluoride and acidulated phosphate fluoride in the toothpaste.

6.16.5. Daily Requirement

Not clearly known but 0.25–0.5 mg per day is considered satisfactory.

6.16.6. Adverse Effects

Excess fluoride damages teeth. This is particularly serious when teeth are coming up, *i.e.*, from birth to ten years of age. Excess intake of fluoride in the body (fluorosis) may cause calcification of ligaments and tendons and leads to muscle and joint problems.

6.17. Molybdenum

Molybdenum takes part in activity of many enzymes.

6.17.1. Functions

Molybdenum is a metallic *co-factor in many enzymes-systems, e.g.*, in protein metabolism, xanthine oxidase and related enzymes. It promotes growth and protects against some cancers.

6.17.2. Principal Food Sources

Liver, cheese, egg, legumes, nuts, whole grains are good sources of molybdenum.

6.17.3. Chemical Aspects

Molybdenum is a hard metal under Group VI - B of the Periodic Table, having atomic number 42, atomic weight 95.95 and 2, 3, 4, 5, 6 valences.

6.17.4. Deficiency Symptoms

Rarely found. Growth may be retarded in children in molybdenum deficiency.

6.17.5. Daily Requirement

0.045 mg may be satisfactory.

6.17.6. Adverse Effects

Not known

6.18. Chromium

Chromium is involved in glucose and lipid metabolism and is a metallic *co-factor* in many enzymes.

6.18.1. Functions

Chromium acts as a metallic co-factor in many enzyme systems, *e.g.*, in glucose and lipid metabolism, insulin production *etc.* It is associated with metabolism of carbohydrate and fat. The metal has essential role in brain functioning.

6.18.2. Principal Food Sources

Red meat, grape, whole grains, cereals, nuts, molasses, brewer's yeast *etc.* are good sources.

6.18.3. Chemical Aspects

Chromium is a very hard metal, under the Group VI – B of Periodic Table. Its atomic number is 24, atomic weight is 51.996 and with 2, 3, 6 valencies.

6.18.4. Deficiency Symptoms

Deficiency symptoms of chromium are usually not observed.

6.18.5. Daily Requirement

0.035 mg may be daily requirement for an adult (= estimated basis).

6.18.6. Adverse Effect

Excess chromium in the body is toxic. It gives disagreeable taste in mouth, diarrhoea, cramp and abdominal pain.

6.19. Bromine

Bromine remains in association with chlorides.

6.19.1. Functions

Bromine as bromide form helps in activating the enzyme, pancreatic amylase. It also helps in activating ptyalin secretion. Bromide is also a central nervous system depressant.

6.19.2. Principal Food Sources

Sea fishes, sea weeds *etc.* are good sources. Bromine is also available from meat, fish, egg and some cereals.

6.19.3. Chemical Aspects

Bromine is a liquid, non-metallic element. It does not occur naturally as an element and exists in combined state. The element comes under the Group VII – A of the periodic table and has atomic number 35, atomic weight 79.904 and 1, 3, 5, 7 valencies.

6.19.4. Deficiency Symptoms

Not clearly known.

6.19.5. Daily Requirement

Only traces are required.

6.19.6. Adverse Effects

Excess of bromine in the body is toxic. It results in vomiting, respiratory trouble, corrosion of oral and gastro-intestinal tract and other health hazards. (Addiction to bromide is observed in some persons which should be discouraged).

6.20. Boron

Boron is of great help for bones and muscles.

6.20.1. Functions

Boron makes effective utilization of calcium and magnesium in the body. It is needed for development of bones and muscles. In preventing body aches and arthritis, it is useful.

6.20.2. Principal Food Sources

Boron is present in many fruits and nuts.

6.20.3. Chemical Aspects

Boron is a hard, non-metallic solid, which is found only as a compound, *e.g.*, boric acid and borax. It comes under Group III – A of the Periodic Table. Atomic number of the element is 5, atomic weight 10.811 and valency as 3.

6.20.4. Deficiency Symptoms

Not clearly known. Weak bones and osteoporosis may result in deficiency.

6.20.5. Daily Requirement

Only traces are required.

6.20.6. Adverse Effects

Excess boron in the body causes nausea, vomiting, diarrhoea, nervous depression *etc.*

6.21. Macro- and Micro-Mineral Elements of some Common Food-Stuffs

Macro- and micro-mineral elements as stated by Gopalan *et al.* in their book, "Nutritive Value of Indian Foods," (2018) revised by Narasinga Rao *et al.* (Published by the National Institute of Nutrition, Hyderabad, under the Indian Council of Medical Research, N. Delhi) have been enlisted in Table 15. The amount for some selected food-stuffs have only been presented. For the amount present in other food-stuffs, the original title may be consulted.

Table 15. Macro- and Micro-minerals Content in some Food-stuffs under different Groups

Food-stuff	*Macro- and Micro-elements*												
	**A*	*B*	*C*	*D*	*E*	*F*	*G*	*H*	*I*	*J*	*K*	*L*	*M*
							****I.**						
a	42	296	8.0	137	10.9	307	1.06	1.15	0.07	3.1	0.02	147	39
b	26	215	1.67	21	-	-	1.19	1.03	-	1.2	0.02	130	91
c	25	222	4.1	171	7.3	131	0.46	0.78	0.04	1.6	0.01	54	44
d	10	280	2.8	157	-	-	0.24	1.10	0.08	1.4	0.01	-	-
e	9	143	1.0	61	-	-	0.17	0.66	0.05	1.3	0.01	-	-
f	10	190	3.2	-	-	-	-	-	-	-	-	-	-
g	10	160	0.7	64	-	-	0.07	0.51	0.05	1.3	-	-	-
h	67	1410	35.0	-	-	-	-	-	-	-	-	-	-
i	48	355	4.9	132	20.0	315	0.51	2.29	0.04	2.2	0.01	122	29
j	23	121	2.7	54	9.3	130	0.21	0.62	0.01	0.6	-	115	47
							II.						
a	56	331	5.3	130	73.2	720	1.34	1.05	0.12	1.7	-	160	39
b	154	385	3.8	130	39.8	800	0.93	0.96	0.43	3.0	0.03	174	9
c	75	405	3.9	122	27.2	1150	0.39	1.02	0.45	2.8	0.01	214	25
d	73	304	27	90	28.5	1104	1.20	0.69	0.28	0.9	-	177	5
							III.						
a	39	44	0.8	31	-	-	0.02	0.18	0.08	0.3	-	-	-
b	830	57	0.9	44	-	-	0.10	0.15	-	0.2	0.01	81	198
c	50	28	2.4	30	58.0	33	0.08	-	-	-	-	27	23
d	73	21	1.1	64	58.5	206	1.10	0.56	0.01	0.3	0.01	30	54
							IV.						
a	18.3	55	1.2	9	59.8	43	0.29	1.05	-	0.91	0.01	14	24
b	46.9	50	0.6	16	4.0	127	0.18	0.96	0.03	0.41	0.01	-	-
c	46	50	0.2	27	9.0	393	0.22	1.02	-	0.11	0.01	-	-
d	10	40	0.5	30	11.0	247	0.13	0.69	0.07	0.53	0.01	37	16

Food-stuff	Macro- and Micro-elements												
	*A	B	C	D	E	F	G	H	I	J	K	L	M
V.													
a	18	47	0.4	15	3.0	200	0.12	0.13	–	0.22	0.01	44	52
b	20	10	0.5	26	1.8	87	0.03	0.06	–	0.22	0.05	10	8
c	33	57	1.2	18	53	138	0.13	0.10	–	0.40	–	231	34
d	66	56	0.4	53	6.9	103	0.11	0.15	–	0.42	0.01	30	41
e	10	30	0.4	38	5.6	139	0.05	0.05	–	0.26	0.01	16	4
VI.													
a	230	490	5.1	373	–	–	0.97	1.88	–	3.57	0.16	–	–
b	50	450	5.8	349	–	–	1.66	1.42	–	5.99	0.16	–	–
c	10	240	1.7	–	–	–	–	–	–	–	–	–	–
d	140	430	7.7	–	–	–	–	–	–	–	–	–	–
e	100	380	2.6	302	–	–	1.67	2.62	–	2.32	0.10	–	–
VII.													
a	30	80	4.4	272	–	–	1.4	1.4	0.07	1.8	0.04	–	–
b	150	282	68	278	–	–	0.4	0.4	–	2.7	0.07	–	–
VIII.													
a	50	20	1.2	–	5.0	225	–	–	–	–	–	–	–
b	10	14	0.7	7	28	75	0.1	0.1	–	0.1	0.01	7	1
c	85	50	0.6	–	–	600	0.2	–	–	–	–	–	–
d	17	36	0.4	41	37	88	0.2	0.2	–	0.2	–	7	8
e	20	23	0.5	82	–	–	0.2	0.1	–	0.1	0.01	–	–
f	10	28	0.3	24	5.5	91	0.1	0.1	–	0.2	0.01	14	4
g	10	35	0.7	10	125	159	0.3	–	–	–	–	19	3
h	14	16	4.3	270	26	205	0.1	0.1	–	0.3	0.01	17	3
i	17	13	0.5	11	6.0	69	0.2	–	–	–	–	13	11
IX.													
a	480	350	3.1	–	66.0	173	0.11	–	–	–	–	–	–
b	180	280	2.1	–	52.0	183	0.14	–	–	–	–	–	–
c	530	235	0.9	–	50.0	151	0.12	–	–	–	–	–	–
d	650	175	1.0	13	101	288	0.13	–	–	–	–	103	3
X.													
a	70	260	2.5	–	–	–	–	–	–	–	–	–	–
b	60	220	2.1	–	–	–	–	–	–	–	–	–	–
c	25	245	–	–	–	–	–	–	–	–	–	–	–
d	150	150	2.5	–	33	270	–	–	–	–	–	–	–

Food-stuff	*Macro- and Micro-elements*												
	**A*	*B*	*C*	*D*	*E*	*F*	*G*	*H*	*I*	*J*	*K*	*L*	*M*
XI.													
a	210	130	0.2	–	19	90	–	–	–	–	–	–	–
b	120	90	0.2	–	73	140	–	–	–	–	–	–	–
c	170	120	0.3	–	11	110	–	–	–	–	–	–	–
d	28	11	–	–	32	130	–	–	–	–	–	–	–

*A: Calcium milligram per 100 gram of edible matter. B: Phosphorus milligram per 100 gram of edible matter. C: Iron milligram per 100 gram of edible matter. D: Magnesium milligram per 100 gram of edible matter. E: Sodium milligram per 100 gram of edible matter. F: Potassium milligram per 100 gram of edible matter. G: Copper milligram per 100 gram of edible matter. H: Manganese milligram per 100 gram of edible matter. I: Molybdenum milligram per 100 gram of edible matter. J: Zinc milligram per 100 gram of edible matter. K: Chromium milligram per 100 gram of edible matter. L: Sulphur milligram per 100 gram of edible matter. M: Chlorine milligram per 100 gram of edible matter.

****I: Cereal grains –** a – Bajra. b – Barley. c – Jowar. d – Rice, parboiled, handpounded. e – Rice, parboiled, milled. f – Rice, raw, handpounded. g – Rice, raw, milled. g – Rice bran. i – Wheat flour, whole. j – Wheat flour, refined. **II: Pulses –** a – Bengal gram, daal. b – Black gram, daal. c – Green gram, daal. d – Red gram, daal. **III: Leafy vegetables –** a – Cabbage. b – Curry leaves. c – Lettuce. d – Spinach. **IV: Roots and tubers –** a – Beet root. b – Onion big. c – Sweet potato. d – Potato. **V: Other vegetables –** a – Brinjal. b – Bottle gourd. c – Cauliflower. d – Okra. e – Pumpkin. **VI: Nuts –** a – Almond. b – Cashewnut. c – Coconut, fresh. d – Pistachio nut. e – Walnut. **VII: Spices and condiments –** a – Chilli, green. b – Turmeric. **VIII: Fruits –** a – Emblica. b – Apple. c – Bael. d – Banana. e – Grape, blue variety. f – Gauva. g – Litchi. h – Mango, ripe. i – Papaya, ripe. **IX: Fishes** – a – Bhekti, fresh. b – Hilsa. c – Katla. d – Rohu. **X: Meat. fowl and eggs –** a – Eggs duck. c – Egg, hen. c – Fowl. d – Mutton, muscle. **XI: Milk –** a – Milk, buffalo's. b – Milk, cow's. c – Milk, goat's. d – Milk, human.

Chapter 7

Water

The principal constituent in the body of human beings, all animals and all plants is water. It is the *most vital and most abundant component* in an organism.

A human being can survive upto 4 to 6 weeks without taking food but it is impossible to survive for more than 2 or 3 days without taking water either as plain water or as fluid in any other form. That is, death will result if there is a loss of 20 per cent of body weight by deprivation of water in the body that takes place in 2 to 3 days.

All fluids in the body, such as blood, lymph and other tissues, all secretions, such as salivary and gastric juice, bile and sweat and all excretions, such as urine and faeces contain water. In a normal healthy individual weighing around 70 kg, total body water constitutes 45 to 55 litres and this is about 10 per cent less in women.

In the human body, water is present both inside the cells (= intracellular fluid) and outside the cells (= extracellular fluid). The water which exists inside the cells constitutes 50 per cent of body weight and that which is present outside the cells comprises 30 per cent of body weight.

There is no part of the body which does not contain water but the amount may be high or low or very low. Thus, as high as 80 per cent water or even more is present in the nervous tissue while as low as 2 per cent water is present in the enamel content of the teeth. (Enamel is a white coating of the crown of the teeth and this is the *hardest substance* in the body).

Human beings consider water as the cheapest of all foods. But a person who has ever experienced what thirst is, understands the value of water. Drinking water has its real respect where it is not available. ("Remember: "Water, water everywhere. Nor any drop to drink," Samuel Taylor Coleridge). In some deserts and hill-stations having scarcity of water, water-carriage conveyed by regularly train or motor van brings a sigh of relief to the inhabitants. "Cold water should thus, be thrown on them who waste this valuable gift of the Mother Earth".

7.1. Water Content of Body Parts

Water content in the body depends on a variety of factors, such as age, gender, body parts, health, activity, climate *etc*. Some people have the habit of drinking too much water and some people drink relatively less amount. In accordance with, water content in the different parts of the body differs.

A report provided by NutrientsReview.com has presented the following Table (Internet) as regards water content of the human body according to age.

Table 16. Water Content in the Human Body According to Age

Age (Body Weight)	*Average Total Body Water (TBW)*
Newborn (6.6 lbs or 3 kg)	75 per cent (2.2 litres)
One year (22 lbs or 10 kg)	60 per cent (6.5 litres)
Male : 19 – 50 years (154 lbs or 70 kg)	60 per cent (42 litres)
Male : 51+ years (154 lbs or 70 kg)	55 per cent (38.5 litres)
Female : 19 – 50 years (143 lbs or 65 kg)	50 per cent (32.5 litres)
Female : 51+ years (143 lbs or 65 kg)	45 per cent (29 litres)
Obese adult (220 lbs or 100 kg)	45 per cent (45 litres)

Water content in the body is known to vary according to gender. There are several reports which state that in adult male, the average water content is around 60 per cent and in adult female, the average water content is around 55 per cent.

There has been a great variation in water content in regard to the organs and tissues of the human body. H. H. Mitchell has stated (Journal of Biological Chemistry, 158) that 73 per cent water is contained by brain and heart, 83 per cent in lungs, 64 per cent in skin, 79 per cent in muscles and kidneys and 31 per cent in bones.

C. C. Chatterjee in his classic book, "Human Physiology : Volume 1." (2004) (Published. by Medical Allied Agency, Kolkata - 700 009) has presented the percentage of water in different organs. According to him, percentage of water stands as 20 in skin,

75–80 in muscles, 76 in blood, 92 in blood plasma, 60 in connective tissues, 60 in corpuscles, in nervous tissue : 85 in grey matter and 70 in white matter. Besides, in adipose tissue, 20, in dentine, 10, in bones (without marrow) 20 and in cerebrospinal fluid 99 per cent respectively.

The said author opined that the figures presented were approximate and average. It has been pointed out by him that the water content of tissues and organs differed from time to time in accordance with loss of water and its provision while degree of activity could also play part. The author went on to comment that the body actually used to get water from two sources : from *food and drink* and as *end-products of metabolism*. According to him, the water received by the body from food and drink should be called *exogenous water* and that water obtained internally as metabolic end-product should be termed as *endogenous water*. The different food-stuffs give rise to different amount of endogenous water. It has been stated that 100 gram of fat, starch, protein and alcohol could respectively produce 107, 55, 41 and 117 gram of metabolic water.

Extracellular and intracellular liquid have drawn much attention of the physiologists. Extracellular liquid has been grouped as transcellular water, water of connective tissue and cartilage, plasma water, interstitial liquid, lymph and bone water which is inaccessible. Transcellular fluid has been considered to include cerebrospinal fluid, synovial fluid, intra-occult fluid and fluid of the pleura, pericardium and some other organs.

7.2. Functions of Water in the Body

Life cannot be conceived of without water. Our Mother Earth contains water where biological organisms exist but this is absent in other planets for which they do not allow any organism to live there. There are plentiful functions of water in the human body which are not possible to mention and some of the pertinent ones have only been listed in the following:

(i) Water brings innumerable constituents into soluble form. Hence, in the science of chemistry, it is regarded as a *universal solvent*.

(ii) It is an ideal medium for many *physical processes*, notably osmosis, diffusion, filtration *etc.*

(iii) Water has high *dielectric constant* for which oppositely charged ions exist together.

(iv) Acids, bases and salts are *ionized* in water which on reaching optimum concentrations are ulilized in the body.

(v) *Metabolic reactions* are not possible in the body without water.

(vi) Water makes *softening of food* and thus, helps in swallowing them.

(vii) It maintains proper concentration of *hydrochloric acid in the stomach* for digestion. *Digestion of food* in the gastro-intestinal tract and *absorption* of nutrients from the intestines cannot take place without adequate water in the gut.

(viii) Flow of *blood, lymph, fluid* around brain, spinal cord, gastric and pancreatic juice takes place through water as the medium.

(ix) Presence of high water in the *cerebrospinal fluid* saves brain and the nerves.

(x) *Lung cells* are kept moist by water.

(xi) *Transport* of nutrients and other water-soluble materials from one part of the body to the other part is accomplished by water.

(xii) Various *secretions* in the body are maintained in proper concentration by water.

(xiii) Removal of waste materials from the body in the form of urine, faeces, saliva, tear *etc.* are carried on by water.

(xiv) Necessity of water has been there for *lubricating* joints, conjunctiva, exposed parts, sex organs and other parts of the body.

(xv) Water is beyond doubt, an excellent medium to regulate *body temperature.* It does the function of regulating body temperature in the following ways. (a) Specific heat of water is high and unlike other fluids, it takes up more heat to raise temperature, (b) it is very efficient in transfer of heat from one part to the other and (c) it makes great loss of heat through urination. Thus, it is an excellent material to protect the body a gains hotness.

(xvi) Water prevents hardening of stool and helps in easy and smooth evacuation.

(xvii) Drinking of water before meals is believed to help in losing weight.

(xviii) Pliability of skin and luster of hair and skin are maintained by adequate water intake.

7.3. Effects of Water Deficit in Body

Existence of certain amount of water is indispensable for the human body or any organism to survive. However, under certain conditions, water content in the body may go lower than the optimum level. This may occur due to (a) external cause or (b) internal cause. More common external factors responsible for lowering of body water may be stated as high atmospheric temperature that prevails and heavy physical exercise undertaken by a person. High loss of water from the body that may take place on taking diuretic drugs (which are usually prescribed in hypertension, oedema, or other conditions) may also lessen body water level. It may also be stated that some people habitually drink less water and cause lessening of body water, although reverse is also true.

Internal factors causing lowered level of body water may be stated as high and frequent vomiting, diarrhoea, some abnormal condition like polyurea (frequent urination) which may particularly occur in children due to diabetes insipidus and other unhealthy condition. Postpartum diabetes may also be a condition for high urination. This occurs in women shortly after childbirth due to rapid fall in oestrogen level.

Some common symptoms of insufficient water in the body are headache, loss of appetite, decreased volume of urine, discolouration of nails, skin abnormality, seizures *etc.* Formation of kidney stone, pathogenic infection of urinary tract organs and menstrual irregularity may also result.

Low water content upto a certain level is tolerated by the body. But if it becomes high, serious effects may be produced. It has been on record that decrease in body water from 5 to 8 per cent may cause dizzines and fatigue, loss over 10 per cent may bring about physical and mental deterioration with unbearable thirst and *if dehydration occurs upto* 20 per cent of body water, it may be fatal and may even lead to dealth of a person.

Regulated starvation may have some benefit in the body but stoppage of drinking water with starvation may be extremely harmful. In some parts of India, it is advised by some ignorant and elderly ladies in the family to drink less water after child birth. There is no scientific reason of this concept and it may be extremely dangerous to adopt this practice.

7.4. Water Requirement in the Body

How much water should a person drink every day is a question which is difficult to be answered precisely. This is because, a host of factors are associated with it and some of the pertinent factors may be stated as follows : age of the person, gender, body weight, physical health, physical activity, thirst, volume of urine output, renal functioning, diabetes of type 1 or 2, sweating, atmospheric temperature, relative humidity of the atmosphere, disease condition, oral rehydration therapy, mental condition, loudness of voice, availability of potable water and so on.

However, many recommendations have been made in regard to the amount of water which is to be taken by a person. Some of these may be stated as follows:

A great many dietitians advise that a person should drink at least 8 ounces of water every day. Many dietitians advise that 1900 to 2000 millilitres of water should be taken by a person who gets 1900 to 2000 calories of energy from food daily. A report of the European Food Safety Authority in 2010 stated that a minimum of 2 litres of water should be consumed daily by a man and 1.6 litres by a woman. It has been suggested by the National Health Service in the U.K. that 1.9 litres of water including that water which remains in food should be taken by a person daily. This amount would be suitable for temperate climate but in hotter climate, the intake should be more.

The reference daily intake for water in the U.S.A. recommended intake of 3.7 litres per day for a man and 2.7 litres for a woman who are older than 18 years. The Institute of Medicine, U.S.A. in 2004 suggested that while a man should consume 3.7 litres of water daily, a woman should consume 2.7 litres.

To suggest how much water should be taken by an adult on daily basis, a formula of 8 into 8 has been put forward by the dietitians in the 1940s. This formula denotes that an adult needs 8 glasses of water in a day and each glass should contain 8 ounces of water. That is, 1920 millilitres of water should be consumed by an adult. However, researches done thereafter have provided evidence that for adequate hydration in the body, a male adult should take 3.7 litres of water daily and a female adult should take 2.7 litres. But the amount should be higher in hot weather, heavy physical exercise, pregnancy and lactation. If other drinks, fruit juices *etc.*, are taken, amount may be lesser. It should not be forgotten that even if thirst is not felt, optimum amount of water must be taken by a person daily in order to maintain hydration of the body.

Alcohol should not be regarded to be a drink for hydration of the body.

7.5. Water Balance

A proper balance is necessary to be maintained in regard to intake of water and removal from the body. In a healthy person, it remains more or less same. This is, if water taken is 2–2.5 litres, its elimination through different channels should be almost same or slightly lesser. This is referred to *proper balance of water* in the body.

But this may so happen that intake in the volume of water in the body may have significant difference with the volume of water that is eliminated. Such a condition is called *improper balance of water*. If the intake of water is more than removal from the body, it is referred to as *positive balance* which indicates that some water remains stored in the body without being removed. This is sometimes observed in children, pregnant women and athletes. On the other hand, if elimination of water is greater than intake, the condition is called *negative balance* of water in the body. Thirst, oedema in the body, consumption of fatty food, administration of diuretic medicine *etc.* may be possible causes for such condition. In some people, fear, anxiety, anger, depression *etc.* may also bring about such condition. Improper balance of water to a little extent is not considered to be of clinical significance unless the differences of intake and removal of water are not differed widely, *i.e.*, unless there is significant difference.

7.6. Adverse Effects of Excess Intake of Water

It is well-known that life cannot exist without intake of water and health is impaired if water intake in the body is not adequate. But it is also true that *excess intake* of water, *i.e.*, much more than requirement makes more harm than any good.

As a matter of fact, many *electrolytes are removed* from the body through urine, perspiration and faeces if excess water is taken. As a result of loss of these ions, body badly suffers. The very common symptoms of loss of ions from the body are fatigue, nausea, muscular cramp, less attentiveness *etc.* Swelling and discolouration of hands, palms, lips and feet may also occur.

Loss of potassium ions to a great extent may very often occur due to too much intake of water. This is called *hypokalemia* which hampers the balance between the intracellular and extracellular potassium ions. Low potassium level causes diarrhoea, prolonged sweating and hampers cardiovascular functioning. *Hyponatremia* is another condition in which high loss of sodium ions takes place and intake of excess water may be a cause of this. Low level of sodium ions in the body causes muscle weakness, weight loss and other abnormalities. Such a condition affects the brain also such as swelling up of the brain, speech defect, walking defect *etc.* Volume of blood may increase due to loss of sodium and potassium ions and other electrolytes. High volume of blood exerts great pressure on heart and lungs.

Excess intake of water has a serious bad effect in that, it *exerts great pressure in the kidneys.* Glomerular filtration rate of a normal healthy person is about 125 –140 ml in a minute but it exceeds when excess water is taken and thereby, exerts much pressure on the kidneys. Stressful condition of the kidneys may cause their damage which may even be irreparable. (Some medical providers advise their patients to take plentiful water in a day. Such an advice without any reason or undesirable symptoms may be of harm to the patients).

Chapter 8

Anti-nutrients

As a coin has two faces, a food has also two opposite sides. That is, a given food may be adequately enriched with components that may have great value in providing nourishment to the body. On the other side, the same food may contain some substance(s) which is liable to counter the effect of the nutritive and other health giving constituents of that food. These undesirable components which are naturally present in a food tend to devalue its quality and nutritive property. This happens so due to the existence of one or more *mineral elements* which act in either retarding or completely inhibiting *absorption* and *utilization* of other constituent(s) that have nourishing, *i.e.*, a limenal value. These are referred to as *anti-nutrients, anti-nutrient factors* or components. The *anti-nutrients therefore, stand in the way* in regard to availability, *i.e.*, utilization of one or more nutrients in the body.

It is pointed out that the anti-nutrients not only disallow some nutrients in making them available to the body but some of these anti-nutrients also tend to inhibit the *activity of the enzymes*. As a result, digestion of food and other metabolic process in the body are also hampered.

The anti-nutrients may be *naturally occurring* in food-stuffs or they may also be synthetic. Chemically, these are various minerals, cations and anions, various forms of flavonoids, proteins, polysaccharides (= glycans) or may be some other compounds.

8.1 Natural Occurrence

Anti-nutrients are found to occur in almost all types of foods although amount of them differs. However, they are of widespread occurrence in *food-stuffs that are of plant-source* than those of the animal-origin foods. It is often noticed that they are usually of higher abundance at the outer side of a plant-source food than in its inner side. In legumes and cereals, relatively high level of anti-nutrients are noted in the hulls than in the grains. Even water and other drinks may also contain anti-nutrients.

It is brought to notice in this connection that as the anti-nutrients are in much higher amount in plant-source foods, consumption of too much vegetarian food and less of non-vegetarian food brings about higher intake of the anti-nutrients in the body. (However this is not always correct).

A number of undesirable symptoms may result in the body due to entry of anti-nutrients and they differ according to the type of anti-nutrients.

8.2 Making Reduction in the Level of Anti-nutrients

Whether and how far, the levels of the different anti-nutrients could be brought down from the various food-stuffs have drawn much attention of the scientists. Cooking in an effective manner is of great benefit to lessen their content or prevent their action. Soaking the seeds, grains or nuts overnight in water and then throwing away the water, spouting, boiling with water and then discarding the water, frying in oil or ghee *etc.* are practical methods to reduce the level of anti-nutrients to a considerable extent.

As regard lessening the level of anti-nutrient, *viz*, phytic acid, some methods are found to be helpful. *Milling and soaking* remove phytic acid to a great level. Removal of phytic acid by the process of *fermentation* could also be done. The optimum pH which is required for enzymatic degradation of phytic acid could be provided by natural fermentation. Another effective method of removal of phytic acid of seeds and grains is to germinate them and then use. It has been reported that as high as *40 per cent removal* of this anti-nutrient could be done only by germination. It has also been on record that addition of *vitamin C* to a food could markedly lower the phytic acid content in it.

Other Uses

Anti-nutrient compounds have however, *some beneficial effects*. Some of them have medicinal value and thus, used in some pharmaceutical preparations. In some chemical industries also, some of these compounds are employed. In food preservation industries, they are used as food additives and preservatives. They play part in growth of the seedlings and act in storage of minerals.

Notable Compounds

There are quite a large number of chemical compounds which are recognized as anti-nutrients and the list is on the increase. It may be mentioned that many synthetic compounds which are of late, increasingly added to cooked or uncooked foods for value addition (= improving quality), storage or other purposes are increasing the list of anti-nutrients. However, some of the more common types of anti-nutrients that have been known may be stated as follows:

(i) Phytate
(ii) Oxalate
(iii) Tannins
(iv) Saponins
(v) Protease inhibitors
(vi) Glucosinolates
(vii) Amylase inhibitors
(viii) Goitrogens
(ix) Arabinoxylans
(x) Gossypol
(xi) Beta-glucans
(xii) Lectins
(xiii) Haemaglutanins
(xiv) Xenobiotics *etc.*

A few of these have been discussed in the following:

(1) Phytates (= Phytic acid esters)

Phytic acid was discovered in 1903 mostly as phytate (in the phytin form) and is also known as inositol hexakisphosphate (IP-6), inositol polyphosphate or as phytate. It is an ester of inositol and phosphate.

Phytic acid has a *strong binding capacity of dietary minerals like calcium, iron, zinc and magnesium*. Hence, it inhibits absorption of these minerals in a body. It also inhibits absorption of the important B-vitamin, *i.e., niacin*. Besides these, phytic acid inhibits digestion of starch, protein and fats at in blocks the activity of the *digestive enzymes*, pepsin, amylase and trypsin. It is on record that zinc deficiency was once a serious situation among the people of Egypt and Iran due to the presence of phytic acid in the diet of the people.

Chemically, phytic acid has 6 phosphate groups and at suitable pH, it is partially ionized. Its chemical formula is $C_6H_{18}O_{24}P_6$, molar mass is 660.029 g, mol^{-1}, E number is E 391 and the IUPAC name is (1 R, 2 S, 3 r, 4 R, 5 S, 6 s) - cyclohexane - 1, 2, 3, 4, 5, 6 – hexayl hexakis [dihydrogen (phosphate)].

Phytic acid is in relatively high amount in the hulls of nuts, cereals, millets and pulse grains. It occurs in many other agricultural and horticultural crops, notably, maize, wheat, rice, barley, sorghum, rye, oat, kidney bean, broad bean, pea, cowpea, lentil, chick pea, soybean, sesame, rapeseed, sunflower, peanut (= ground nut), almond, cashew nut, walnut, hazel nut *etc.* Amount is however, relatively high in rice bran, sunflower, linseed, sesame, soybean *etc.*

In the life of plants, phytic acid has a beneficial effect in that, this is the main storage form of phosphorus in the seeds and is a source of cations and myoinositol.

(2) Oxalate (= Oxalic acid esters)

Oxalate (= Oxalic acid ester) is derived from the weed plant, known as wood-sorrel (*Oxalis* species) and hence, named as oxalic acid.

Oxalic acid is poisonous. On coming into skin contact, blisters are formed and if the chemical goes inside the body, symptoms such as abdominal pain, burns, convulsions, vomiting, shock *etc.* may result and such condition is known as *oxalemia*. Regular intake causes kidney stone formation and accumulation of high oxalic acid content in blood is known as *oxalemia*.

As an anti-nutrient, oxalic acid binds dietary *calcium* and make them into insoluble calcium salts. As a result, the calcium that comes into the body through diet is *not absorbe*d and remains *unavailable* to the body. It is likely that some other minerals apart from calcium may also be formed into insoluble salts by this acid.

Oxalic acid is the simplest form of discarboxylic acid, which is crystalline, colourless and gives a colourless solution in water. It has reducing property and is a *chelate* for metallic cations. Although a dicarboxylic acid, oxalic acid is a *strong acid*. Its chemical formula is $C_2H_2O_4$, molar mass is 90.034 g, mol^{-1} and the systematic IUPAC name is ethanedioic acid.

Oxalic acid occurs in abundance in the weed plants, wood-sorrel (*Oxalis corniculata* L.), *Biophytum sensitivum etc.* and many crop plants, *e.g.*, spinach, rhuburb, parsely, amaranth, carrot, radish, lettuce, turnip, tomato, sweet potato, pea, cucumber, coriander, almond, carambola, bilimbi *etc.* The level is in higher amount in cabbage, cauliflower, brussel's sprout, spinach, rhuburb *etc.*

Oxalic acid is however, used in many industries. It is used as a bleaching agent in pulpwood, in dying, in removal of rust, in making baking powder *etc.*

(3) Tannins

Tannins bind with *iron* and inhibits its absorption in the body. They bind *proteins* as well and make them unavailable to the body. Tannins are largely found in seed coats of pulse seeds and in many fruits and vegetables as well as in millets like ragi, bajra *etc.* They are condensed form of polyphenolic compounds, amorphous, astringent and precipitate gelatin, many alkaloids and glycosides from solutions. They absorb oxygen in alkaline solution.

(4) Goitrogens

These are substances present in some plants which inhibit absorption of *iodine* by the thyroid (= endocrine) gland. Thus, the pituitary is stimulated and thyroid stimulating hormone (TSH) is released. As a result, growth of thyroid tissue is triggered and ultimately *goitre* is formed. Thiocyanate, iso-thiocyanate, glucoinolates are more common of the goirogens. Some food crops like soybean, cauliflower, cabbage, radish, brussel's sprout, turnip, beans *etc.* contain some goitrogens in relatively high amount. Iodine absorption may be hampered by nitrates also.

(5) Saponins

These are substances which if mixed with water and shaken, give rise to *soapy foam* (= *lather*). They bind with *bile salts* and *cholesterol*. Saponins are amphipathic glycosides and are present in a large number of plants and notably those belonging to the plant family, sapindaceae. However, among animals, presence of saponins has been noted in the marine animal, sea-cucumber. Saponins are poisonous. Their presence protects the plants against attack of animal and microbial pests.

(6) Trypsin Inhibitor

Trypsin is an important enzyme which has indispensable role in hydrolysis of proteins in the human gut (Ref. Chapter 3). Trypsin inhibitor is a proteinaceous compound which interferes enzymatic activity of trypsin for which, *digestion of proteins is hampered*. This inhibitor impedes functioning of the *chymotrypsin* as well. The inactive form of trypsin is *trypsinogen* which is formed in the pancreas and this is activated to form trypsin by peptidase enzyme. Trypsin inhibitor results in many intestinal diseases also.

This protease inhibitor is formed in a number of plants, *e.g.*, papaya, jackfruit seed, soybean, pineapple and in many pulses. The activity of trypsin inhibitor is however, destroyed by *heat*. Hence, cooking of foods remove this inhibitor from them either completely or to a considerable extent.

Besides the above, there are many other compounds which inhibit absorption and thereby, bio-availability of mineral elements or beneficial compounds in the body to a greater or lesser extent.

It may be mentioned that the level of any inhibitor in a crop plant depends on certain factors. Varietal difference is observed in some crops such that the amount of inhibitor may be relatively less in some variety(ies) than the amount in other variety(ies). Decrease of some inhibitors with growth of fruits is also observed. For example, tannin content of fruits like guava, pomegranate, sapota *etc.* decrease with development and ripening. Considerable portion of an inhibitor could be removed from a food by soaking in water, boiling, frying and roasting.

Chapter 9

Dietetics

Nutrients are indispensable substances for an individual to sustain life and carry on the daily activities while *food is the carrier* through which these agents are delivered into the body. The primitive people used to satisfy hunger on consumption of the flesh of wild animals and the fruits that were grown in the jungles. When they came to family life, they learned the technique of raising *fruit trees* by sowing seeds which they had learned from the women in their families. They used to take a variety of foods, which were both of animal and plant sources. However, in course of time, they learned to sort out the food items keeping in view their availability, palatability and the benefits that could be obtained to defend against and cure illnesses of various types.

With rolling of ages, it was realized by the ancient people that in order to work effectively, *two full, i.e., square meals in a day at suitable interval of time* were necessary as the main or the principal foods. Although some foods might be taken in between these two principal meals, these should be considered as light meals, which are called *reflections* and hence, these meals should as they considered to the consumed in relatively less amount. The necessity of maintaining time in taking the principal meals during the day and night to facilitate digestion was realized by them, particularly after they acquired the knowledge of cooking. In this way, the concept of taking *regular and organized diets* came to light.

9.1. Terminology

Before going into the study of diet and its impact on health, it seems necessary to become acquainted with the terms that are commonly used in this specialized area of health science and these have been discussed in the following:

(1) Food

Food (derived from the Anglo-Saxon word, *foda*) of an organism in any material that is suitable for its intake into the body, satisfies hunger, provides the nutrients to sustain life, maintains physiological functioning and does not bring about any adverse effect on consumption. For the human beings and animals, food constitutes materials that are derived from *plants and animals and is not an inorganic material except water and salt under the condition of normal health.* The type of food of an organism differs according to species, races, country, region, availability and many other conditions.

(2) Appetite

Appetite is the *desire* to take food. Food is taken *pleasurably* when there is appetite and unwillingly when there is no appetite. Even when there is no hunger, appetite may be stimulated in a person just by seeing, smelling, reading, thinking or talking about any particular food of choice. (Imagine, if a person even thinks of tamarind, his or her mouth cavity might be filled with saliva). Appetite is stimulated due to emotion, stress and remission of fever also.

(3) Hunger

It is a type of sensation that results from lack of food. It is associated with weakness and an *extreme desire, i.e., craving* to eat. Some stimulants trigger hunger, such as empty stomach, heavy physical works undertaken, food taken in the previous meal(s), some hormones *etc.* Some signs of hunger are hunger pangs (= momentary and violent pain), dull pain in the lower part of chest, restlessness, light-headedness *etc.* Hunger is regulated by a part in the hypothalamus of brain which is in the lateral side of it.

Hunger is sometimes confused with appetite but both are not same. It may be stated that *hunger is the need to eat while appetite is the desire to eat.* Hunger is difficult to control in most cases but appetite can be controlled without much difficulty and thus, may even be ignored by a person. When there is real hunger, persons usually cannot wait long for food.

Appetite is in fact, a *pleasurable sensation* and a person with appetite may have a smiling face. On the other hand, a person with hunger may have a rude appearance. If the person is not polite and not mannerly, he or she may even use uncivil words. ("A hungry stomach has no ears.") Thus, hunger is an *uncomfortable sensation* caused by lack of food while appetite is a *pleasant sensation* which is based on previous experience that makes a person to seek for food to relish the food.

(4) Satiety

Satiety is a feeling, that is sensation in a person which denotes that he or she is fully satisfied, in other words, repleted on taking food. Clearly speaking, the person has no desire to take the same food or drink any more. The hunger is controlled in a person by the *lateral hypothalamic area* in his or her brain, while satiety is controlled by the *ventromedial hypothalamic nuclei in the brain*. It may be mentioned in this connection that satiety is not exactly same with satiation and the latter feeling has some difference with the former.

(5) Diet

The term, diet, which is derived from the Greek word, *diaita*, refers to the sum of the solid and liquid forms of foods that are taken by a person. Diet may be of two types which are, *normal* diet and *prescribed* diet.

A normal diet is one which is regularly, *i.e., habitually* taken by a person. There are four conditions in a normal diet. These are, it should (i) satisfy hunger. (ii) palatable to the taste of the person, (iii) provide all the six nutrients, *i.e.*, carbohydrate, protein, fat, vitamins, minerals and water in right amount and (iv) not cause any undesirable effect either in the body or in the mind. The normal or habitual diet is sometimes referred to as dietitian of a person.

A prescribed diet refers to a diet which a person is advised for some specific purpose. Such a diet is a modification of normal diet. Most commonly, a modified, *i.e.*, prescribed diet becomes necessary when the person has any illness or abnormality of health. However, apart from medical ground, a diet may be prescribed for other purposes. This may be for cosmetic purpose, reduction or increase of body weight, economic condition of the person and the like.

(6) Dietetics

The science of dietetics has its root in Greek, *diatetikos*. This refers to the carefully selected solid and liquid foods and proportion of each such food which are to be taken by a person who may be healthy or sick. In case of a healthy person, the diet should aim at improving health. Evidently, such a diet should contain *all the six nutrients* and in proper amount of each and besides that, the diet should also contain required amount of soluble and insoluble forms of *dietary fibres*.

For preparing diet for a sick person, the dietitian should have adequate knowledge on the ailment of the patient as well as knowledge on what diet has its necessity for the ailing person. Accordingly, it becomes necessary for the dietitian, to have consultation with the attending physician of the patient. The process is best done if the dietitian, physician and the patient *sit together* and formulate the diet. The patient should bring to the notice his or her intolerance or hypersensitivity to any food(s) if known to him or her, preference in regard to type of foods, mode of preparation, availability of the foods in the locality and economic condition of him or her.

It should be remembered that diet of a sick person is actually a *part of the treatment* and hence, it should be designed carefully in consideration to many factors. In the recent years, a new area in health science has come into being which is known as *dietotherapy*. This medical speciality is originated from the fact that some ailments could be completely or almost completely cured only by the use of efficacious diet and without taking any medicine.

It is considered nowadays that a dietitian should be both a *nutritionist and dietitian*. He or she must have formal education in these areas in an accredited University or Institution and must have obtained degree or diploma in the subject. Some educational institutes have introduced basic knowledge of anatomy and physiology as well as practical training in the curriculum. It is also mandatory that the dietitian, in other words, the nutrionist-cum-dietitian *must have Govt. registration*.

9.2. Importance

Dietetics has a great role both for healthy body and healthy mind of people. In India, studies and researches on dietetics and nutrition have although must been done during the past few decades, necessity is felt much later for further attention on this science with production of expert dietitians in the country for a number of reasons, among which the following are of significance.

(i) Malnutrition is known to be a serious problem among a considerable part of the population in this country. A scrutiny reveals that both *under-nutrition* and *over-feeding* have been common in the country. The extent of malnutrition in India and the possible causes have been stated in the Chapters 1 and 10. Children's under-nutrition has been more acute in the country where *more than one-third of the under-nourished children live.* Over-nutrition also stands in the way of healthy living of the people and it has been revealed in 2010 that 14 and 18 per cent of women and men of India are respectively afflicted with obesity (corpulence). Evidently, intake of proper diet needs to be ensured among the people to overcome the malnutrition situation.

(ii) For an ailing or a convalescing person, intake of right type of diet as stated above is a part of medical treatment. While giving dietary prescriptions to a patient, his or her choice of food should no doubt, be given due attention. Thus, instead of prescribing only barley water, biscuits and loaves which had been a practice in olden days, foods of other types such a rice-pulse medley (= khichri, khichuri), gruel made from sago, soaked rice flakes (= chira), puffed rice (= muri), parched rice or wheat (= khoi of rice or wheat) which have greater likeness to the Indian people may be advised where there is suitable. In West Bengal, a low-calorie diet is often prescribed which is made from dried and ground kernels of water chestnut (= paniphaler palo) to sick persons. Importance of such food which are more likable to the Indian people needs emphasis.

(iii) Substitution of food items should be heedfully considered where necessary. Accordingly, where mother's milk is not adequate, properly formulated milk for feeding the neonates and children needs to be explored. Substituted diets for the people having intolerance to lactose sugar in milk, gluten and gliadin proteins in wheat and cereals needs exploration.

(iv) Formulation of diets for the children and elderly people of India needs adequate exploration with due attention to the choice of food and dietary habit of the people residing in different parts of the country.

(v) Low cost diets to suit the economic condition of the people belonging to different groups needs adequate attention.

(vi) As a large part of the population in the country refrain from taking animal source food, especially meat, fish and egg, studies on complete and partial vegetation meals need critical attention.

(vii) Food-stuffs that are usually uncommon to the people in the country such as some species of algae, millets, sea fishes, small snails, fruits and nuts of trees grown as wild *etc.* are needed to be critically studied for their utilization as food for the people. Cooking and suitable preparations of these materials to suit the taste of the people are areas that need exploration and thus, culinary science and technology need great attention. (Culinary science and technology has of late, assumed the status of a need-based curriculum of course in schools and colleges).

9.3. Factors Affecting Diet

There has been a plentiful variety of diets that are taken by people living in different parts of the world. The recipes of these diets are made on paying consideration to a number of factors and the more common factors may be stated as follows:

(1) Religious and Ethical Factors

Religious, spiritual or philosophical beliefs play great influence in the regulation of diets of people belonging to the respective groups, particularly in India. These schools of religion made restrictions on some particular foods with the belief that intake of those foods will stand in the way to worship gods and godesses with devotion. Some examples are cited below:

(i) *Hinduism* : In this religion, a number of restrictions have been imposed in regard to intake of food. Hindu religion has several sects and broadly, two cults are prevalent now. These are, *Vaishnavas* and *Shaktas*. For followers of the Vaishnavas, there is rigid restriction in taking any type of food which is of animal origin. For the followers of Shaktas, meat and fish are permitted only on special occasions but intake of egg or meat is strictly forbidden. In most of the groups belonging to the above two sects, onion and garlic are not permissible either in cooked or in uncooked form. Intake of salt is restricted on some particular days in the month. To pay reverence to some gods or goddesses in some temples, the devotees vow not to take any particular fruit throughout their life.

(ii) *Buddhism* : Specific instructions on the intake of food are hardly available in Buddhism but at least the Buddhist monks and the highly devoted

Buddhists take only vegetarian food and no animal source food is taken by them except milk. Taking diet of such type is a mark of their great respect to the first Five Precepts of Lord Buddha. (Ahimsa Parama Dharma).

(iii) *Jainism* : The diets of the Jainas are similar to that of the Buddhists but are much more rigid. Intake of food of animal source is strictly forbidden among the followers of Jainism except milk. Among plant source food-stuffs, root and stem tuber crops that grow underground are not permissible to be taken.

(iv) *Islam* : In Islam religion, only *halal* food is permitted to be taken. Severe restriction has been imposed among the Mohammedans in consumption of any *haram* food, which includes meat of any animal which has not been killed following Islamic method of ritual slaughter. Taking wine is also forbidden among the Mohammendans.

(v) *Kosher* : The set of Jewish dietary rules is said to be Kosher. Some foods and food preparations are recognized as Kosher in Judaism.

Besides the above, people belonging to some sub-groups of the main religious groups maintain their own principles in taking food items in their diet.

(2) Personal Choice

Choice of food by a person is a great factor that greatly determines his or her dietary schedule. Some people in India are found to be fond of non-vegetarian foods like meat and egg while some non-vegetarians have a great preference for fish. On contrary, others have fondness to vegetarian food. It is known that people in some parts of India are accustomed to take milk to a very high amount. Use of soured foods and chilli to a relatively high extent in daily diet is very common among the people in south India while sweet preparations are liked by people in West Bengal and the eastern India.

(3) Availability of Food-stuffs

Diet of persons depends much on availability of food-stuffs in the locality. High consumption of rice and rice-based food products is common in south India, West Bengal and most parts of the eastern States in the country owing to the fact that rice is available there to a great extent unlike wheat and other cereals. On contrary, consumption of wheat is high in northern India for high production. Inclusion of fruits and nuts in the diet is common in Kashmir, Himachal Pradesh and other parts of northern India for their easy availability.

(4) Regional basis

In India, people of vegetarian and non-vegetarian food habit greatly varies according to regions. The report states that (Google) the percentage of vegetarians comes to 63 in Rajasthan, 62 in Haryana, 48 in Punjab, 45 in Gujarat, 35 in Madhya Pradesh and 33 in Uttar Pradesh. Hence, dietary habit of the vegetarian people differs from those of the non-vegetarians in the same States. However, the *vegetarians in India consume milk*, although it is a non-vegetarian food and thus, they should be regarded as *lacto-vegetarians* rather than complete vegetarians.

(5) Food Intolerance

Intolerance to some foods like *lactose in milk, gluten and gliadin proteins in cereals* as stated in Chapters 2 and 3 is a factor that determines diet of people having such physiological abnormality.

(6) Medical Advice

It becomes necessary for many sick persons to take diets of special type which are to be prepared as per advice of the physicians or dietitians as diet is a part of the medical treatment. Accordingly, diets of such people are different from the normal diets. The recipes of a number of medically prescribed diets have been stated under the next heading. Thus, this is a factor that determines the type of food ingredients and their quantity to be included in a diet.

(7) Family Advice

In India, advice given by the elderly people in the family has a marked influence that particularly affects diet of children and sick persons.

(8) Economic Condition

A person's purchasing capacity of food-stuffs in the market is beyond doubt, a pertinent factor that determines the food items and their quantity which are to be included in his or her diet. This is one of the reasons why proverty-stricken people cannot take meat, fish, egg or milk regularly.

9.4. Types of Diets

Hundreds of dietary specifications are prevalent in different parts of the world and the number is always on the increase. Dietitians' perusal on the factors that are associated with dietary schedule as have been laid down in the Chapter 9.3 has evidently been the motivation for so large number of dietary formulations.

Before selecting which diet is to be adopted, a person should however, take into consideration a number of factors and the important ones may be stated as follows:

(i) Despite the fact that many of the formulated diets are of great benefit, many are also of little effect while many may even adversely act upon health. Mention may be done to the Gerson diet (= Gerson therapy) which advises intake of low salt, low fat and thrusting on vegetarian food. The formulation was said to prevent and also to cure cancer and degeneration of health but these were not proven by the American Cancer Society and on contrary, serious illnesses had been observed with this diet.

(ii) Taking any particular diet over a *prolonged period* should be done with great care. This is because, a sick person may go on taking a given diet for some period but if the sickness turns a different direction or if the person is on the way of convalescence, the same diet may not be necessary and may even bring about harmful effect. Hence, regular *health check up* becomes necessary and the person should always get in touch with the nutritionist-cum-dietitian and the attending physician as long as the diet is continued. For example, a hypertensive person who takes low or no extra salt in the diet for a long period may develop very low blood pressure and in such situation, if the same diet is continued, it may prove to be fatal and may cause comatose. Hence, occasional master health check-up cannot be looked down upon by a person.

(iii) Self imposition of a dietary schedule without adequate knowledge may bring about serious health hazard. Sometimes friends of a person having no adequate knowledge come to impart dietary advice. If a person really seeks to know what diet has to be taken by him or her and how long that should be continued, he or she should consult a proficient dietitian and it is still better to consult a nutritionist-cum-dietitian to get the real benefit of diet. It should be remembered that dietetics is a science and an integral part of health science. Those who do not have adequate background on dietetics should not give recommendation to a person for it may be highly dangerous. (This is a bad habit of many Indians).

(iv) Health of one individual is not exactly same with another. So also a person afflicted with a particular disease may not have same situation with another person who is attacked with the same disease. In accordance with, medicines of the two persons may not be alike and diet should also differ. This is an aspect which is given adequate attention in the Homoeopathic system of medicine.

(v) In formulating scientific diets, myths and household recommendations should neither be over-emphasized nor totally disregarded. It may be stated that years back when small-pox was a serious disease in India, many physicians used to advise diets to the patients as recommended by the elderly members of the family for they had learnt to treat patients from their past experiences. Giving foods prepared with pure ghee was advised by the elderly ladies to the patients suffering from some types of morbidness and particularly, respiratory illnesses.

(vi) Diet needs serious consideration for the pregnant and lactating women. They should take diet as per advice of the gynaecologists, paeditritians or expert nutritionists-cum-dietitians. It is also necessary that they should be in regular touch with the health experts as the prescribed diets may have to be changed with time in accordance with health and clinical condition of them.

(vii) A person having any congenital health abnormality, immunity disorder, intolerane to foods like lactose (= milk sugar), gluten or gliadin protein of cereals or hypersentitivity (= allergy) to any food should be brought to the notice of the dietitian while seeking dietary advice. Other health abnormalities like irritable bowel syndrome, existence of gastric or duodenal ulcers, inflammation of appendix or prostrate, renal calculus, diverticulosis, chronic costiveness *etc.* should be made known to the dietitian.

(viii) It is essential for the dietitian to know whether the person is a smoker or has any habit of using tobacco (khaini, zarda, snuff *etc.*) in any form or addiction to alcohol or sleeping droughts. Even psychic behaviour of a person should not also be set at defiance in prescribing diet.

(ix) A person seeking dietary advice should not hesitate to speak to the dietitian whether vegetarian or non-vegetarian foods are liked by him or her and even the culinary process of them.

Dietary Prescriptions

It has been stated above that *plentiful types of diets* exist around the world. Most of these have been formulated either to avert or to cure some diseases, although in many specifications, the health has not received much attention.

In accordance with the objectives to ward off diseases or to cure against them, some of the common dietary prescription have only been enlisted in the following:

(1) Cardiovascular Health

There has been much attention in the recent times to formulate diets that are not harmful to cardiovascular health. Such diets have been designed with two objectives, which are, diets to aim at maintaining healthy cardiovascular functioning and diets to be taken after cardiac myopathy, *i.e.*, abnormality, if any, to protect the cardiovascular functioning without further deterioration.

In the light of the fact that cardiovascular dysfunctioning takes effect for some abnormalities notably, hypertension, hyperglycaemia, hypercholesterolemia, hypertriglyceridemia *etc.*, diets which are efficacious to retard in developing these conditions are framed so as to maintain healthy functioning of the cardiovascular system. Some of these diets have been stated in the subsequent part.

Dietarian, *i.e.*, food habits of the people living in Mediterranean countries is considered by many dietitians to be of much benefit for cardiovascular health. In this connection, "French Paradox" needs pointing out. Many people in France are of the opinion that although the French people consume high amount of butter, cheese, cake and other food that have high amount of saturated fat, which are liable to being about atherosclerosis and thereby, cardiac abnormality, cardiovascular complications are much less prevalent in them as compared to the people of many countries. According to them, the French people *drink too much of red wine* and as this wine contains a stilbenoid polyphenol compound known as resveratrol, they are not easily afflicted with cardiac abnormalities. Led by this concept, Prof. Serge Renaud of the Bordeaux University in France introduced the term, "French Paradox" in 1991.

However, it needs mentioning that the above concept was *not supported by all*. Even the World Health Organization commented upon, "Neither the food of the French people nor the red wine had any influence on cardiac efficiency. It is also true that the magnitude of heart disease among people in France is no less than those of the people in the neighbouring countries". However, the French people consume too much fruits, and this is no doubt, of advantage for healthy functioning of the cardiovascular system. For heart diseases, a *prudent diet* designed by te American Dietetic Association is known to be of benefit.

Diets designed after a cardiac myopathy take into consideration a number of factors. These are, diets should not be such that they tend to raise the level of cholesterol, particularly LDL-cholesterol, triglyceride and sugar in the blood and should not also increase the blood pressure. Besides these, diets should not produce hyperacidity, distention and constipation. Soft diets, or at times even liquid diets are usually formulated intelligently for coronary patients to avoid food related health abnormality.

(2) Hypertension

All dietetians and cardiologists advise that people having high blood pressure (as regular occurrence) *must restrict intake of salt*. Complete absence of salt (= common salt) in the diet is perhaps impractical but the intake should be low to very low depending on the extent of blood pressure According to many, it should not exceed 500 mg in a day. It should be remembered that salt remains as salt whether it is taken without cooking or after cooking. Instead of common salt, *i.e.*, sodium chloride, potassium chloride is advised.

A diet known as DASH diet, which is an abbreviated form of *Dietary Approaches to Stop Hypertention* promoted by the National Heart, Lung and Blood Institute in the U.S.A. gets much favour to control hypertension, *i.e.*, high blood pressure. The speciality of this diet is, nuts, fruits, grains, vegetables and chicken should be in much higher amount in the diet while sweets and red meat should be in very low amount and above all, intake of salt should be extremely low.

Kempner rice-fruit diet is another diet to lower hypertension. This diet formulated by Walter Kempner in U.S.A. make a rigid restriction of salt and recommends rice, fruit and sugar such that the body does not get more than 2000 kcal and 7 mEq of sodium per day.

(3) Hypercholesterolemia

In order to lower the cholesterol level in the blood stream, rigid restriction should be done on the foods having high amount of saturated fat, such as butter, cream, lard, bacon, red meat, pork, coconut, palm oil *etc*. On the other hand, foods having higher amount of *mono-unsaturated fats, e.g.*, almond, hazel-nut, cashewnut, pecan nut, pictachio nut and fruits like olive, avocado, guava, citruses *etc*. should be taken in higher amount. It is advisable to cook food in canola oil, oilve oil, groundnut oil, sunflower oil *etc*.

It should be noted in this connection that foods having higher amount of polyunsaturated fats such as walnut, corn, pumpkin *etc.* although reduce the level of cholesterol but such foods lower the good cholesterol (HDL) level also and hence, are of inferior standard. Poly-unsaturated fats are high in oil of cotton seed, soybean, safflower, margarine *etc.* and hence, cooking with these oils lower the good cholesterol level in blood.

(4) Arresting Blood Clotting

Clumping together of blood cells (= platelets) to form clot(s) in the blood stream is a serious health hazard that leads to *thrombosis.* Foods rich in omega-3-fatty acids are found to be of help in preventing the blood cells from clotting. Among other foods, fishes like mackerel, tuna, salmon and hilsa are of much help to the patients who have suffered from thrombosis. Cooking foods in soybean oil and canola oil is also advised to lessen the chance of thrombus formation in blood.

(5) Hyperglycaemia

Persons having diabetes mellitus whether of type 1 or type 2 are first of all advised to *restrict intake of sugar* which include monosaccharides and disaccharides. They should occasionally get the blood sugar tested, both after fasting and post-prandial and also glycosylated haemoglobin and make sure of their blood sugar level. Intake of sugar and the amount if any should be done as would advise by the physician on examining the reports and the medicine, if any taken. Intake of *refined carbohydrates* should be restricted by them in accordance with medical advice.

A diet plan known as *Zone diet* renders benefit to the diabetic persons. This diet advises that the carbohydrate, fat and protein in each meal should not exceed 40, 30 and 30 per cent respectively. Besides this, body weight should also be controlled.

South Beach diet is another dietary formulation which centres primary attention on consumption of unrefined and slow carbohydrates.

A diet popularly known as *Atkins diet* claims to be of benefit to the diabetic persons. This diet suggests that foods which are to be taken should be low in carbohydrate but high in protein. However, possible risk of Atkins diet has been put forward by some dietitians.

Some food therapists are of the opinion that some particular food-stuffs are of much benefit to control blood sugar level among the diabetics. Daily intake of unsweetened yoghurt is stated to be of great help in this respect. Under Indian condition, curd which is soured milk and is known as *dadhi, dahi, doi etc.* also reduce blood sugar if taken without addition of sugar. But it should be noted that curd made from some particular species of the bacterial genus *Lactobacillus* is actually of great worth and not from all species of *Lacobacillus.*

Among fruits and nuts, walnut, almond, blueberries, grapefruit *etc.* are also recommended. Sweet orange juice especially of the fruits of the Mosambi variety is claimed to be of much benefit whose glycaemic index score is 40. Vegetables like tomato, beans, kale *etc.* and some grains, notably bengal gram, lentil, barley *etc.* have been stated to reduce blood sugar level to a considerable extent. Effect of some fishes like sardine, mackerel, hilsa, herring to bring down blood sugar level has been brought to notice by some food therapists. Gopalan *et al.* (2004) from the National Institute of Nutrition, Hyderabad have stated in their book, "Nutritive Value of Indian Foods" that seeds of fenugreek, *i.e., methi* (*Trigonella foenum graecum*) have significant role to diminish the level of blood sugar.

It is therefore, worthwhile to suggest that the above items of food should be included in the diets of persons whose blood sugar level is high, *i.e.*, in hyperglycaemic condition.

(6) Obesity

Obesity, in other words, corpulence has now assumed a grave situation among the Indian people. It has become an "unrecognized and dangerous epidemic in India" and is more acute among the urban people. The exact cause of obesity is yet to be known but certain factors play great influence.

Diet has undoubtedly a *great impact to prevent obesity as well as to redress it* to a considerable extent in a person, who is declared to be an obese person. However, before discussion about the anti-obesity diet, it is necessary to have an understanding on two aspects, which are, what is really meant by obesity and what are the undesirable effects of it.

When a fat person especially a boy or a girl is met with, many people have the tendency to say, "You have become obese, so make reduction of your food." It should be noted that addition of fat in the body does not necessarily mean obesity and making curtailment of food without medical advice may be dangerous. Obesity is in fact, assessed on examining the "Body Mass Index" of a person. How to determine the Body Mass Index (BMI) has been described under the heading, 1.4.4 in Chapter 1.

Although BMI is an accepted formula to assess obesity but there are some limitations in the method as it considers *only height and weight of a person and not waist.* According to Prof. Gita Mathai of the Christian Medical College in Vellore, India, the BMI of the Indian should be "ideally around 23 and preferably under 25." She went on to state that "waist circumference should be about 102 cm in men and 88 cm in women. The waist-hip ratio (circumference of the waist divided by that of the hips) should be less than 0.9 for men and 0.85 for women."

If deposition of fat is considered to determine obesity, it should be remembered that not only outward accumulation but deposition of fat in the internal organs is far more dangerous. Fatty liver is a serious abnormality of the liver. Hardening of arteries due to deposition of cholesterol and other materials results in high blood pressure and thus, a major cause of cardiac abnormality known as atherosclerosis. Among other factors, gender is known to play part as it is stated that *women deposit fat under the skin while men deposit inside veins and arteries.* Hence, atherosclerosis is more common in men.

Obesity is not only a problem of cosmetic importance but it is responsible for a number of serious diseases. Resistance to insulin leading to hyperglycaemia is very often associated with obesity (= corpulence). Other health hazards are hypertension, cardiovascular diseases including atherosclerosis, arrythmia, venous thromboembolism, heart failure *etc.* Deranged lipid profile, polycystic ovary syndrome, Crushing's syndrome, early onset of osteoarthritis, Alzheimer's disease and even cancer are also known to be associated with obesity.

(An interesting study on the cause of obesity has been recently done by Prof. Qi Chen and his team at the University of California at Riverside, California in the U.S.A. They observed that men who took poor diets produced "damaged sperms which put their children at risk of obesity and diabetes". Both these diseases run in families and were earlier believed to pass into generations through DNAs only. But Prof. Chen and associates brought forward that not only DNAs but RNAs also transmitted hereditary properties to the offspring).

Principles of Dieting for Obsessed Person

The following principles should be kept in consideration as regard diet to be taken by him or her.

(i) Energy, which is measured in terms of Calorie (or Joule) should not be stored much in the body. The foods that go inside the body give rise to energy. The energy is expended by physical activity and also in functioning of internal organs, *e.g.*, respiration, blood circulation *etc.* (known as basal or resting energy) if the energy that is obtained from foods is *far more than what is expended,* the extra energy that remains in the body will give rise to production of fat, effectuating obesity.

As stated in previous Chapters, different foods give rise to different amount of energy. That is, 1 gram of carbohydrate or protein obtained from food gives rise to 4 kcal (= 16.8 KJ) while 1 gram of fat obtained gives 9 kcal (= 37.8 KJ) of energy. Now the energy which gets into the body through food should be in consistence with the energy expended. Thus, dietary regime of a person who undertake heavy physical work should not be same with that of a person who does mostly sedentary works or carries out moderate physical works.

(ii) Modification on the consumption of foods to reduce fatness should be done in consideration to their *energy giving value only.* Hence, great care should be ensured that the *body is not deprived to get other nutrients through foods.*

(iii) Curtailment in amount of food should not be done which denotes that the stomach should be fulfilled. Fulfillment of the stomach should in fact, be done by consumption of much of *dietary fibres.* These roughage foods as stated under the heading, 2.8 (b) in Chapter 2 do not give rise to significant energy in the body but fulfills the stomach and have many merits. Thus, the motto should be "Eat more and reduce."

(iv) During the period of taking up an anti-obesity diet, regular check up of the body in regard to BMI is necessary.

(v) Dietary advice or prescription should be obtained only from a govt. registered nutritionist-cum-dietitian. During the course of taking the prescribed diet, if the person feels any physical discomfort, illness, fatigueness, loss of weight or depression, he or she should bring that to the notice of the dietitian without delay.

Recommended Diets for Obsessed Persons

A number of diets have been prescribed by dietitians all over the world. Many food companies also market a number of specially formulated food products. Some organized diets have been stated in the following:

Montignac Diet is a type of diet which is based on the concept that the carbohydrates that are to be taken should only be such which have low glycaemic index. *Monotropic Diet* is one which is designed with the principle that only one type of food should be taken during a period until loss of weight has taken place to the desired level. *Okinawa Diet* is a Japanese diet which recommends consumption of such foods only that give rise to low energy and is designed, keeping in view the food habits of the Ryukyu island.

A crash diet, known as *Subway Diet* has popularity which consists of some special type of sandwich formulated food designed by commercial concern. Popularity of *Pritikini Diet* which is a Japanese diet is very high. This diet suggests that foods should be cooked with plenty of soured milk and very little oil. *Yo-yo* diet is one which is based on the principle of alternating cycles of losing and regaining weight and thus, it is also known as *Weight Cycling Diet*. The *Jenny Craig Diet* is a type of pre-packaged food made by the Jenny Craig concern.

Fruits have of late, received much attention in formulating diets of obese persons. Grapefruit diet is one (Grapefruit is a large type of citrus fruit, botanically named as *Citrus paradisi* Macf., juice of which is slightly bitter due to the presence of bitter principle known as naringenin) suggests intake of high amount of grapefruit juice in the diet. *Beverley Diet* (Beverley is an important fruit growing zone in the U.S.A.) pays emphasis on taking high amount of fruits in the diet to lose weight.

To lose weight in a short time, partial starvation is recommended by some dietitians. One such diet is known as *One Time Meal in a Day*, which is abbreviated as OTMD. This diet suggests that an obese person should take only one instead of two principal meals in a day.

There are many such diets which are advised by cosmetic-purpose dietitians. The principles of these diets are, reduction in consumption of food and reduction in accumulation of calorie in the body. As regard reduction in consumption of foods, it may be opined that it may cause harmful effect such as high accumulation of acid in the stomach, constipation *etc.* Attempts to prevent calorie intake by reducing food intake point out the fact that, even if calorie through consumption of food is restricted, the body knows how to obtain calorie from the glycogen which is already stored in liver, muscles and some other organs.

(7) Improper Growth of Children

Before making a special diet plan for a child for the purpose of increasing growth, at first it becomes necessary to confirm from a nutritionist-cum-dietitian and also from a child specialist physician whether the child really needs it. It should be noted that particularly in the urban areas, quite a number of children are bulked with food which results in obesity in them. (Parents of many children are of the wrong opinion that fat children are more healthy !)

(8) Intolerance to some Foods

It has been stated earlier that some people are not tolerant to some chemical compounds which are present in some types of foods. Some examples are, *gluten* and *gliadin* present in wheat and some cereals that cause the disease, *coeliac sprue* if breads made from their dough are taken, lactose sugar in milk that causes *lactose intolerance* (Chapter 3, Heading 3.15 a) *etc.* These aspects are needed to be carefully considered in formulating diets.

(9) Gastrointestinal Disorders

For persons suffering from gastrointestinal disorders, *e.g.*, indigestion, distention, malabsorption, irritable bowel syndrome *etc.*, *Elimination Diet* is necessary to be followed which identifies food(s) that causes adverse effect to a person by the process of elimination, one by one. Swallowing cold milk in low amount at short intervals is advised to a person suffering from intestinal ulcer. In amoebic hepatitis, *High Protein Diet* is advised by many physicians. This diet is considered helpful for the problem of intestinal worms also. Inclusion of adequate amount of insoluble dietary fibres in the diet is advised to persons suffering from chronic constipation.

(10) Inability to Processed Food

A diet known as *Macrobiotic Diet* is of late, gaining importance in which processed pulses and grains are excluded. Inclusion of some sea weeds after boiling or cooking is done in this diet and sometimes little amount of ghee, butter or margarin is advised in this diet. Some physicians are of the opinion that Macrobiotic diet may be of help to arrest spread of cancerous cells in cancer patients. In the *Pritikini Diet* also, processed foods are eliminated,

(11) Difficulty in Chewing

For persons having difficulty in chewing, *Soft Diet* is recommended. These diets are made by mashing rice, boiled potato, carrot *etc.* and making a mixture of them.

(12) Renal Complication

Persons having renal impairment should take diet which is low in protein. A dietary schedule known as *National Renal Diet* has been designed by the American

Dietetic Association and the National Kidney Foundation of the U.S.A. which consists of six food planning system. Dietary restrictions are to be rigidly followed by the patients undergoing dialysis.

(13) Neurological Disorders

A diet known as *MIND Diet* is formulated which is often recommended to reduce Alzheimer's disease. This diet is designed making a combination of DASH diet and the Mediterranean diet.

At the present time, a diet, known as *ketogenic diet* (Syn. Keto diet) has received much importance in the U.S.A. and some other countries. This dietary formulation has been stated to be of help not only in recurrent epileptic seizures but also in many other abnormalities, notably diabetes, hypertension, hypercholesterolimia *etc.*

The ketogenic diet has some peculiarities in that, it aims to *provide 70 per cent of calories in the body by consumption of fat, less than 10 per cent from carbohydrate and less than 20 per cent from protein.* Thus, the diet should contain high amount of butter, cheese, egg, bacon, salmon, nuts and among plant-source foods, olive oil and non-starchy foods like spinach, cabbage, cauliflower *etc.*

The idea of this diet is to develop *ketosis, i.e.,* accumulation of ketones in the body, such as acetone, beta-hydroxy butyric acid and aceto-acidic acid. (Ketone bodies consist of carbonyl group attached to two carbon atoms).

The definite advantage of ketogenic diet which aims at making a drastic reduction of carbohydrate is yet to be properly established. However, in many cases, it gives *quick result* particularly in reducing blood sugar level.

Many demerits of this diet have been known. *Keto-flu* is a syndrome which brings about severe fatigue, dizziness, loss of sleep *etc.* Constipation, vomiting, hypoglycaemia and greater frequency of epileptic seizures may also take place if the keto diet is not taken in a regulated manner. A keto diet may increase the risk of renal calculus as well. (As a matter of fact, importance of adequate carbohydrate in the diet could not be oversighted).

(14) Weight Gain

In the case of severe loss of body weight, a diet known as *Weight-Gain Diet* is recommended by many dietitians. In this diet, high amount of proximate food components, *i.e.,* carbohydrate, protein and fat is recommended to take. Such a dietary regime should actually be followed for a considerable period of time to get the desired result. Regular check-up of body weight before and during the intake of the diet is no doubt, necessary and care should be taken that the weight gain does not go to super-optimum level.

(15) Muscle Development

To prevent wasting and enhance muscular growth, *High Protein Diet* is most commonly advisable which consists of high amount of first class protein.

(16) Pregnancy and Lactation

Extreme precaution should be taken in formulating diets for pregnant women and lactating mothers. Diets of them (i) must contain all the six nutrients and in proper amount, (ii) should be easily digestible, (iii) facilitate bowel movement, (iv) do not cause hyperacidity, flatulence or vomiting, (v) contain adequate amount of dietary fibres, (vi) should be healthy, free from pathogenic organisms and harmful contaminants and (vii) should be palatable. In case of gestational diabetes, sugar intake should beyond doubt be regulated. Some foods which contain high protease enzymes, *e.g.*, outer side of unripe papaya fruits which contain papain, pineapple which contains bromelain, fig fruit which contains ficin, kiwi fruit which contains actinidin should be avoided. (Folic acid has a great role in child bearing age. See Heading 5:5.6.7. Folic acid may even be given through salt, as is done in iodine).

Some women have the habit of *pica* during pregnancy. This is a peculiar desire to eat some non-food material only to get pleasure. The materials may be of diverse types such as burnt clay, ash, pulses, cereal grains and so on. Although these foods are usually not very harmful, care should be taken that such a pica which causes harm to the pregnant woman or the foetus should never be eaten.

(17) Incidence of Gout

Gout is a metabolic disease, which is mostly hereditary and physiological cause is deposition of uric acid in blood in very high level. It is an acute and common type of arthritis which produces severe pain in any joint of the body and associated with redness and swelling around the joint. Gout results due to deposition of uric acid in excessive amount in blood. Uric acid is crystalline and occurs as end-products of purine metabolism, *i.e.*, when purine is enzymatically broken down. As stated under the Chapter 3 on protein synthesis, uric acid is formed from purine bases which are derived from nucleic acids (DNA and RNA). The crystals of the sodium esters of uric acid, *i.e., ureates deposit* in and around the joint and gives pain.

Some amount of uric acid always remains present in blood but its excessive presence which causes gout depends on a number of factors and among these, food has a great influence. Foods that prevent uric acid to deposit in high levels are yogurt or curd (dadhi) which is not sweetened, milk of low fat content, fresh fruits, nuts, vegetables, potato, mushroom, various grains and particularly whole grains. An anti-gout diet should therefore, include the above food items. Losing weight is an effective way to relieve of gout but it should be slow and in moderation. Atkins diet is not recommended for this purpose and too much loss of weight may be a serious hazard.

Some foods such as red meat, pork, venison, organ meat like liver, kidney, thymus and marine fishes like sardine, tuna, cod, herring, food containing much of saturated fat *etc.* are known to increase the uric acid level and hence, these foods if taken should be in very low amount. As alcohol also raises the uric acid level, it should be discarded or taken in moderation.

(18) Personal Choice

It has earlier been emphasized that choice of food of the person should be given adequate importance in prescribing his or her diet. The principal author has an experience in treating some rural people living in the Bankura district of West Bengal. In case of their diarrhoea, they were advised specific diets. But the people in this region cannot live without rice. Besides, they are also fond of opium seeds (locally known as *posto*). On hearing these facts from them, the principal author advised to take thoroughly boiled rice after repeatedly washing in cold water to remove much of starch. As adjunct, they were advised to take with rice pasted opium seeds adding water but without frying (known as *kancha posto*) and soup of small fish such as *mourola* without adding spice. The patients were advised to take this diet but only for few days and it had worked well. (Long-term use of opium seeds is although common among Bengalees in some parts, this is however, not desirable as the seeds contain some opium although in low amount and brings about constipation as well).

(19) Detoxification

To protect the body against toxic chemicals, a diet known as *Detox Diet* is recommended, in which synthetic chemicals, preservatives *etc.* are not added. High amount of fruits and high water are included in such a diet. However, chemical toxification of the body from agricultural pollution is difficult to be eliminated and organically grown food crops should be used for this purpose, if available.

(20) Post-surgical Care

After surgical operations, *Post-oprative Diets* are recommended. These diets vary according to surgical operations undertaken, age and other factors and aim at providing necessary nutrients to the patient and restoring normal health.

(21) Old Age

Diet is a very important consideration for the elderly persons. This has been discussed later.

9.5. Mixed Diet

In designing diets of normal, healthy persons, it is desirable to make *combination of many types of food*. This is because, if a particular food is deficient in one or more of the nutrients, it is possible to be supplemented by other food(s).

In the Chapter 3 (Heading 3.14), it has been mentioned that taking a diet which is a mixture of rice, wheat and other cereals with that of pulses is of great benefit. The reason is, the cereals are deficient of the amino acid, *lysine* but are good sources of the amino acid, *methionine*. On the other hand, pulses are rich sources of lysine but have deficiency of methionine. So, if a food is prepared by making a combination of cereal and pulse, *deficiency of one or more amino acids may be supplemented by the pulse.*

Pulse preparation, which is taken alone with cereal food in India can be enriched in their protein quality when a mixture of different pulses are used. Protein derived from pulses is usually of lower quality than that obtained from animal source protein. But the quality of protein of pulse preparations can be made improved by making a combination of different pulses while cooking it. Many types of combinations of foods to design diets are made all over the world. These are made by mixing different types of plant source foods, animal source foods and largely by making combinations of both plant source and animal source food-stuffs. Acharya, K.T. has stated various food-stuffs that are made use of in making many food preparations in India. (Ref. Acharya, K.T., 2009, *Your Food and You*, National Book Trust, Govt. of India, Vasant Kunj, New Delhi - 110 070, India).

9.6. Cooking

Cooking was innovated by the human beings after fire was discovered but processing of food was known to them even earlier when they learned the benefit and technique of sun drying. For the human beings of the present day, cooking is almost inevitable for cereals, millets and oilseeds while many of the foods like fruits, vegetables, condiments and spices are consumed after cooking or without cooking, depending on the type of food, maturity condition of the food-stuffs, necessity, preservation, preference by the consumers and on other conditions.

All types of cooking that are practised are of two types, which are *dry method* and *wet method*. Dry method of cooking consists of roasting and wet method involves heating in water or in oil or ghee. Roasting involves drying of processed or unprocessed food-stuffs without use of liquid substance as the media. This involves drying in open sun or on oven and sometimes heating is done mixing with sand. In wet method of cooking, water or oil, ghee *etc.* must have to be used as the medium of cooking. When water is used, it is called boiling and the term, frying is used when heating with oil or fatty material. Both roasting (= drying) and frying may be done to a lesser or greater degree depending on the food-stuffs and choice of the consumers. There are both merits and demerits of drying and frying as stated in the following:

Merits

(i) Palatability, appearance and flavour are most important attributes of cooked foods. Some food-stuffs such as almost all cereals, millets, oilseeds and some pulses, vegetables and fruits are almost impossible to consume without roasting, boiling or frying.

(ii) Heating or cooking has necessity to make the foods chewable. This is of particular importance to the infants, children, elderly people and those who lost teeth.

(iii) Digestibility of many foods is improved by cooking. Digestibility of some pulses and oilseeds like groundnut, Bengal gram, safflower, maize grains, besides potato, brinjal *etc.* is greatly improved even by roasting only.

(iv) Many of the harmful micro-organisms that are present in food-stuffs are destroyed or inactivated by heating or cooking. For some micro-organisms, high temperature becomes necessary for destruction. Cooking temperature may be less if pressure is high and hence, cooking in pressure-cooker is much effective. Cooking time is much less and destruction of microbes is high when cooking is done in a pressue-cooker or a steam sterilizer.

(v) Trypsin enzyme which is present in the intestine acts in digestion of protein (Chapter 3). Some chemical compounds inactivated the action of proteins and *these are known as trypsin inhibitors.* Trypsin inhibitors are found in seeds of some pulses, notably soybean, some cereals, jackfruit seeds *etc.* These are also found in the *white portion of the eggs of ducks.* Protein digestibility is reduced if the trypsin inhibitors go inside the body through foods. However, cooking in high heat for longer time destroys these inhibitors.

(vi) During cooking, the starch grains swell and the cell-wall in them burst. As a result, the enzymes in the digestive tract are easily accessible to the broken starch grains, for which they become easily digestible.

(vii) Mineral nutrients and some vitamins, *e.g.,* carotene (= precursor of vitamin-A) of the food are not lost in cooking by heat. Vitamins are usually saved from destruction from foods when little of citric acid juice or crystal, vinegar, tamarind *etc.* are added during cooking with heat.

Demerits

(i) Some of the proteins in vegetables are lost when cooking is done using salt.

(ii) Many vitamins and mineral nutrients, *e.g.,* sodium, potassium and calcium of the food are lost by leaching. To recover these, leached out nutrients in water may be re-used instead of throwing away. Cooking with less amount of water is of benefit in this respect which makes lesser loss of nutrients by leaching.

(iii) Many of the vitamins which are heat-labile, i.e., their activity is destroyed when heated (Chapter 5) are lost on cooking and vitamin-C is a classic example. As stated under (vii), acidifying the food during cooking may save many of the vitamins to a considerable extent during cooking with heat.

(iv) Repeated heating of cooked food makes a great loss of vitamin-C in them.

(v) Making small pieces of vegetables in cooking brings about greater loss of vitamins, particularly vitamin-C. Hence, making larger pieces is advisable.

(vi) Peeling vegetables to cook make a great loss of vitamins for many vegetables although not in potato. This is because, in many vegetables and most of the fruits, vitamins and minerals are usually in higher amount towards the outer surface and the peel. Peel of some vegetables, *e.g.*, beet, carrot, raddish, corn *etc.* contain some substances which protect them against loss of many nutrients during cooking. Hence, it is advisable to cook them along with peels and the peels may be removed if necessary, after cooking.

(vii) When cooking is done in high heat, the amino acid, *lysine* of protein in the food is combined with some carbohydrates and thereby, it becomes unavailable.

(viii) Quality of protein in milk is deteriorated when boiled for long time.

(ix) By cooking in high heat, many of the proteins in food are coagulated.

It may be stated that some dietitians advise to take some foods without cooking. For example, it is sometimes advised to take milk, beating it with raw egg. This can never be regarded to be a good food and may cause health hazards. In olden days, some physicians used to prescribe raw liver of goat to eat on crushing it to the patients afflicted with anaemia. It should be noted that liver in uncooked form is not only indigestible but can also invite health hazard. Hence, the organ, liver is though of benefit in anaemia, it should be taken after cooking only with heat.

(An interesting and also queerish fact may be brought to light in this connection. All members of a family residing in the Murshidabad district of West Bengal in India are known who habitually take only uncooked, unroasted and even unboiled foods. On interrogation, they replied that they relish to take vegetables as raw and had no liking for cooking food except boiled rice. They emphatically coommented that they had no digestive trouble of any kind on taking raw food. The process of their taking meals had been filmed and movie telecast from the Kolkata Television Centre).

9.7. Vegetarianism

Over the ages, human beings have developed a time-tested food habit and they have started taking both vegetarian foods and non-vegetarian foods. Ancient people used to take both vegetarian as well as non-vegetarian foods but in course of time, they were divided into two groups. One group used to take foods derived only from plant origin while the other group formed the habit of taking foods derived from plant source as well as those from animal source. With passage of time, again they formed groups such that, one group began to take foods derived from plant source but with inclusion of milk which is though obtained from animal source. Those who used to take both plant source as well as animal source foods including milk did not change their food habits.

However, the following four types of food habits are now found to be prevalent among the people in different parts of the world. (i) Those who take such foods which are obtained *only from plants* and not from any animal. (ii) Those who take plant source foods but *include milk* which is though a food of animal origin. (iii) Those who take plant source foods but include some *selected animal source foods* like fish, meat, egg and milk in occasions. (iv) Those who freely take *both plant source as well as animal source foods* without much restrictions.

It is apparent that only the first group of people should be recognized as those of *strictly vegetarian food habit* and their percentage is no doubt very low. Indian people, especially those belonging to the Hindu religion as also the Buddhists however, consider milk as a vegetarian food and following that principle, people who consume *plant source foods as well as milk are recognized as vegetarians.* In that sense, it has been estimated that 23–37 per cent of the people in India are considered as vegetarians.

In India, people who refrain from taking fish, meat and egg on religious ground are on the concept that killing animals for use as food is a cruelty to animals and barbarous as plenty of many foods could be obtained from plants. Those with non-vegetarian food habits have the argument that the human beings brought the animals from the forests and domesticated them to obtain meat and egg. They introduced fishes in the ponds. With ages, the wild animals evolved to form new species or races but they remain to be dependent on the human beings for survival. The human beings rear them only to obtain food. Now, if they stop taking of meat and egg, will they rear the pet animals any more? In that case, will the reared animals go back to the jungles and survive there or the poultry birds will start flying in the sky? The ecologists are also of the opinion that if more and more people adopt vegetarian food habit, domestic animals will increase in number and this will have a bad impact on ecological system.

Vegetarian Diets

Although broadly termed as vegetarian diets, there are many *modifications* of such diets that are adopted by people around the world. A few are stated in the following:

(i) *Vegan Diet*: In this diet, only plant source foods are used and any type of animal source food and even milk are completely avoided. Thus, this diet could be regarded as *strictly vegetarian diet.*

(ii) *Lacto-vegetarian Diet:* Non-vegetarian food is totally discarded in this diet except *milk and milk products* (Lacto = milk). In India, this diet is much popular and is regarded as a vegetarian diet.

(iii) *Semi-vegetarian Diet:* This is a vegetarian diet but *occasionally meat is included in this diet, under certain conditions but not regularly.*

(iv) *Fruitarian Diet*: Fruits of various types form as the predominant part in this diet. Although this is a vegetarian diet but meat, fish and egg are sometimes added to it. This diet is usually advised to a person having deficiency of vitamins and minerals.

(v) *Pescetarian Diet:* This is a vegetarian diet except that fish is used in it but meat of any animal is excluded.

(vi) *Ovo-vegetarian Diet:* In this diet, vegetarian food *along with egg is included* but milk and meat are not used.

(vii) *Ovo-lacto Vegetarian Diet:* Plant source (= vegetarian) food along with *egg and milk are included* in this diet but meat is not used.

(viii) *Pollotarian Diet:* Along with vegetarian food-stuffs, egg and chicken are used in this diet but *meat of any mammal is prohibited.*

(ix) *Pollo-pescetarian Diet*: This is similar to the Pollotarian Diet but fish is used and meat of any mammal is excluded.

(x) *Kangatarian Diet*: This diet is sometimes taken by the Australian people which includes vegetarian foods but meat of the animal, kangaroo is occasionally added.

(xi) *Milk diet:* This diet is recommended by some ayurveds in India. The diet, which is also called *milk therapy* consists mostly milk and milk products is recommended to patients suffering from gastric or duodenal ulcers. The diet is recommended to continue only for a few days and not for long time.

(xii) The principal author of this book recommends a special type of *lacto-vegetarian* diet for the hyperglycaemic and obese persons. Rice should not be included in this diet and instead of that, hand-made breads (roti, ruti) should be taken both in day and night. These are to be made by thorough mixing of 60, 10, 10, 5 and 5 parts of flour, respectively of whole wheat, jowar, bajra, fenugreek (methi) and hushked Bengal gram with required amount water to make the dough. The diet should include adequate boiled bitter gourd. Besides adequate amount of seasonal fruits, skimmed milk (containing less than 0.5 per cent fat), curd, little amount of salt (preferably, sea-salt) and juice of sour lime should be included in the diet.

Regular check up of body weight, body mass index, blood sugar (including glycosylated haemoglobin) is necessary to be done during the period of the above dieting. If advised by the dietitian, the dietary schedule may be partially modified by addition of rice, sugar *etc.* and then discontinued after a week or so.

9.8. Fasting

According to many dietitians, "Fasting is an inseparable part of dieting". This suggests that to maintain health, not only regular intake of food and drink but their *occasional abstinence has also its necessity*. The practice of fasting is in vogue in many countries since ancient time.

9.8.1. Link with Religion

It is interesting that in all religions, fasting is linked with religious practice. There may be two reasons for this. These are, firstly worshipping could be performed in a still better way by the act of regulated fasting. Secondly, it is linked with religion in order that the worshippers can practise it religiously, *i.e.*, strictly.

In the Hindu religion, fasting is considered mandatory at least in the Shivaratri and Janmastami festivals. Hindu religion has many sects and in all sects, fasting *on the eleventh day of the lunar fortnight* (= ekadashee tithi) is religiously advised. This day comes twice in a month, *i.e.*, in the light fortnight (= shukla pakhsha) and in the dark fortnight (= krishna pakhsha). It is stated in the holy scripture in the Hindu religion that on the eleventh day in both fortnights, the human elements are not adequately secreted and this badly affects vitality due to the ill effects of the moon. However, if fasting is done on the eleventh day, the body is saved from the ill effects of moon and vitality is restored.

In the Christianity, eighteen types of fasting have been mentioned in the holy Bible. In the Islam religion, fasting is rigidly practised. In this religion, it is called *roja*, the literal meaning of which is, *to burn off.* That is, in the holy Quaran in the Islam religion, it is stated that the evils of the mind are burnt off by the act of fasting. To practise roja, no food should be taken from sunrise to the sunset but could be taken before or after that. In the Islam religion, six types of fasting have been advised. These are, Ramdan, Monthly, Muharram, Shab-e-Barat, Shawal and Asura.

9.8.2. Fasting and Starvation

Fasting and starvation are not same. Fasting is done for some specific objectives and these may be religious festivals, rituals, ceremonies, customs, observances, respects, venerations, medical advice, clinical examinations *etc.* Besides that, fasting is done for a short period.

Starvation is also a fasting process but it is carried on in a more *severe manner.* A person may starve for reasons like unavailability of food and drink, poverty, as a mark of protest to a motion (= hunger-strike !) or is forced to do so as an act of punishment.

9.8.3. Types of Fasting

Fasting may be of many types. These may be as follows (i) Complete fasting, in which solid or liquid food of any type is not taken. Some fasters belonging to the Hindu religion even refrain from swallowing saliva of the mouth for day and night during the Shivaratri festival. (ii) Solid food fasting, in which solid food is not taken but any type of liquid food may be drunk. (iii) Half-day fasting, in which fasting is done upto midday or thereafter. (iv) Abstinence of taking rice and salt on the day of the full moon and the day of the new moon in order to avert the pain of gout. (v) *Intermittent fasting, i.e.*, skipping food for 12 to 16 hours between two principal meals or taking only water or liquid in between is advocated by many dietitians as it helps *autophagy* and protect against fatty liver and Alzheimer's disease.

9.8.4. Benefits of Fasting

Benefits of regulated fasting have been recommended by many dietitians and Ayurveds. Researches done in Louisiana, U.S.A. provided evidence that skipping one meal in a day for a considerable period and preferably for a year had reduced body weight of persons by 10 per cent besides that, their mood had improved.

Experiments in the U.S.A. have also revealed that pancreas activity of people had improved by fasting.

However, although it may seem that fasting makes a person quite weak and less active, it is observed that it does not happen so if regulated fasting in done for a short period. This is because, the glycogen that remains stored in the liver, muscles *etc.* in the body is a highly rich source of glucose and during fasting, this stored glycogen is enzymatically broken down (= catabolized) to release glucose molecules which are utilized to produce energy. Autophagy (= autophagia) is another advantage. This refers to *self-consumption by a cell.* That is, when fasting is done, the healthy cells of the body do not get food and hence, *they eat away the unhealthy cells as food.* The unhealthy cells cause many diseases but as these are eaten away by the healthy cells, the body is saved from any diseases. This discovery was made by Yoshinori Ohsumi for which he was awarded the prestigious Nobel Prize in 2016.

9.8.5. Adverse Effects of Fasting

Fasting has the greatest demerit in that, it is liable to bring about *dehydration in the body* and this may be extreme if severe fasting is done, particularly without taking water or any fluid. The people belonging to the Islam religion have a good practice that they drink adequate water before they start fasting to observe Ramdan.

Other adverse effects of fasting are extreme fatigue, hyperacidity, heart burn, disturbance of sleep, high loss of ketone bodies and nitrogen through urine and faeces *etc.* Less attentiveness, change in mood, poor memory *etc.* may also be the symptoms. It is advisable for a person to break the fast as soon as such symptoms appear.

9.8.6. Precautions

The following aspects should be kept in mind before, during and after fasting:

(i) Fasting is not recommended to everybody. It should not be done by pregnant and lactating women, elderly persons, persons who do heavy physical works and are sick. It may cause extreme harm to people who have low blood pressure or any type of cardiovascular or cerebrospinal disorders, urinary disturbances, intestinal ulcers, weakness and fever.

(ii) Fasting without taking water or any other fluid is extremely dangerous which may bring about severe dehydration and even comatose condition.

(iii) Heavy physical works should be completely stopped during fasting.

(iv) The faster should not stay in a smoky atmosphere.

(v) Loudness of sound from microphone, fire-crackers *etc.* cause harm to the faster. He or she should also talk less.

(vi) Restrain is a very important factor for the faster and emotional disturbances should be avoided during fasting.

(vii) Just after breaking the fast, only liquid substances such as juice of sweet fruits, cold milk, water of tender coconut, solutions of D-glucose, electrolytes should be taken by the faster.

(viii) It is a very important consideration that during fasting, if the faster has any type of physical discomfort such as extreme fatigue, abdominal distention, scanty urination, headache, nausea, vertigo, high body temperature or mental confusion, fast should be broken forthwith by drinking liquid materials and preferably glucose solutions.

9.9. Dietary Recommendations for the Indian People

What type of diet should be much befitting to the Indian people have drawn considerable attention of many dietitians in different parts of the country Nevertheless, it may be commented upon that any single recommendation is actually difficult to provide for the people all over India as a widely different food habits that are adopted by the people in different part of the country. The topic on dietary recommendations in India has been discussed under the following sub-headings.

The first report on dietetics in India is known to be presented by Dr. Chunilal Bose in Calcutta. This physician, professor and chemist of the Medical College, Calcutta who is regarded as the "Father of Dietetics and Nutrition in India" from his studies had pointed out in his 1917 address to the Science Convention, "the pit-falls of a rice-based diet - different in protein or muscle-forming constituent and over-rich in carbohydrate elements - and suggested corrections that were effective and inexpensive". He emphasized on protein and "advocated adding gram (= Bengal gram) and coconut in vegetable curries for additional protein and fat". (Ref. Chittabrata Palit, "Chemist, Doctor and Patriot," *The Telegraph Daily – Knowhow*, March 12. 2018, Monday, Calcutta-700 001). Subsequently, Acharya Prafulla Chandra Roy, the Palit Professor of Chemistry of the Calcutta University and many other nutritional scientists for different parts of India paid high emphasis in studying and carrying out researches in the area of nutrition and dietetics. Many of these scientists had put forward a number of dietary recommendations for the people of this country.

However, with due attention to the recommendation put forward by the League of Nations in 1937, well-balanced and practical recommendation of dietary allowances of energy, protein, iron, calcium, vitamin A, thiamine, ascorbic acid and vitamin D for the Indian people had been provided by the Nutrition Advisory Committee of the Indian Research Fund Association in 1944. On the basis of habitual diet of the Indian people, a specific "balanced diet" was also formulated to provide the dietary allowance which could supply all the nutrients that could be considered necessary for the Indian people. In consideration to the requirement of protein and energy, the dietary recommendation of 1944 had been revised by the Indian Council of Medical Research in 1958.

The recommendations made by the Indian Council of Medical Research (Govt. of India) were revised after every decade. Thus. the recommendation given in 1958 was revised by the Nutritional Expert Group of the Indian Council of Medical

Research in 1968 in consideration to the nutrients except energy and accordingly, fresh recommendation was given on the dietary requirement. But a separate expert group of the Indian Council of Medical Research reviewed the situation and brought forward a recommendation in 1978, making modification of some of the nutrients included in the recommendation of 1968. Nutrient allowances for the Indian people were again reviewed by an Expert Committee constituted by the Indian Council of Medical Research in 1988. This committee provided detailed recommendations in consideration to the safety levels of each nutrient. The committee suggested the following principles.

(i) Body weight of 60 and 50 kg has to be considered as reference, respectively for man and woman. (ii) Energy value is to be expressed by BMI making a reduction of 5 per cent. (iii) Protein requirement for adult person should be 1 gram per kg of body weight. (iv) For fat requirement, total invisible fat of the cereal food consumed and the minimum requirement of the Essential Fatty Acids need consideration. (v) Iron requirement should depend on absorption level of 5, 3 and 2 per cent. Calcium and phosphorus should be of the ratio 1 and for infants, 1 and parts respectively. Suitable allowances should be there for micronutrients. (vi) Beta-carotene should be 600 microgram. Allowances for thiamine, riboflavin and niacin should be 0.5, 0.6 and 6.6 milligram per 1000 kilocalorie of energy respectively. (vii) Folate requirement should be 3 microgram per kg of body weight and for vitamin B-12, 12 microgram per day, for vitamin C, 10–25 milligram per day and for vitamin E, 0.8 milligram per gram of polyunsaturated fatty acids respectively.

The dietary recommendations for the Indian people which have been provided by the National Institute of Nutrition (Indian Council of Medical Research), Hyderabad have been stated in Table 17. (Ref. Gopalan, C. *et al.*, revised by Narasinga Rao, B S. *et al.*, *Nutritive Value of Indian Foods*, 2018, published by the National Institute of Nutrition, Hyderabad - 500 007) under the Govt. of India.

9.10. Balanced Diet

Balanced diet of a person is a dietary regime which contains *required amount of all the six nutrients* such that the person maintains good health and carry on the daily activities effectively. At the present time however, soluble and the insoluble dietary fibres are also included in the daily diet of a person.

It may be stated in this connection that if the diet is considered to be a balanced diet, *it should be specific to a person*. This signifies that the balanced diet designed for a particular person should not be exactly same with that which is designed for another person. A number of factors in fact, depend in formulating balanced diets and the more important ones may be stated as (i) gender, (ii) age, (iii) occupation, (iv) pregnancy, (v) lactation, (vi) food habit, (vii) existing illness if any, (viii) preference to foods, (ix) psychic behaviour *etc.* Other factors such as availability of food-stuffs, cooking facility and economic status of the person are also notable factors.

Table 17. Recommended Dietary Allowance for Indians (1989)

Group	Particulars	*1	2	3	4	5	6	7		8	9	10	11	12	13	14
								15	16							
Man	Sedentary work	60	2425							1-2	1-4	16				
	Moderate work		2875	60	20	400	28	600	2400	1-4	1-6	18	2-0	40	100	1
	Heavy work		3800							1-6	1-9	21				
Woman	Sedentary work	50	1875							0-9	1-1	12				
	Moderate work		2225	50	20	400	30	600	2400	1-1	1-3	14	2-0	40	100	1
	Heavy work		2925							1-2	1-5	16				
	Pregnant woman	50	+300	+15	30	1000	38	600	2400	+0.2	+0.2	+2	2.5	40	100	1
	Lactating woman															
Infants	0-6 months		+500	+25						+0.3	+0.3	+4				
		50			45	1000	30	950	3800				2.5	80	150	1-5
	6-12 months		+400	+18						+0.2	+0.2	+3				
	0-6 months	5.4	108/kg	2.05/ kg						55ug/ kg	65ug/ kg	710ug/ kg	0.1			
						500								25	25	0.2
	6-12 months	8.6	98/kg	1.65/ kg				350	1200	50ug/ kg	60ug/ kg	650ug/ kg				
	1-3 years	12.2	1240	22			12	400		0.6	0.7	8	0.4	30		
Children									1600							
	4-6 years	19.0	1690	30	25	400	18	400		0.9	1.0	11	0.9	40	40	0.2-1.0
	7-9 years	26.9	1950	41			26	600	2400	1.0	1.2	13	1.6		60	
	10-12 years	35.4	2190	54			34			1.1	1.3	15				
					22	600		600	2400				1.6	40	70	0.2-1.0

Group	*Particulars*	**1*	*2*	*3*	*4*	*5*	*6*	*7*		*8*	*9*	*10*	*11*	*12*	*13*	*14*
								15	*16*							
Boys	10-12 years	31.5	1970	57			19			1.0	1.2	13				
Girls	13-15 years	47.8	2450	70			41			1.2	1.5	16				
					22	600		600	2400				2.0	40	100	0.2-1.0
Boys	13-15 years	46.7	2060	65			28			1.9	1.2	14				
	16-18 years	57.1	2640	78			50			1.3	1.6	17				
					32	500		600	2400				2.0	40	100	0.2-1.0
Girls	16-18 years	49.9	2060	63			30			1.0	1.2	14				

*1: Body wt. Net (kg); 2: Energy (Kcal/d); 3: Protein (g/d). 4: Fat (g/d). 5: Calcium (mg/d). 6: Iron (ug/d); 7: Vitamin A; 8: Thiamin (ug/d); 9: Riboflavin (ug/d); 10: Nicotinic acid (ug/d). 11: Pyridoxine (ug/d). 12: Ascorbic acid (ug/d). 13: Folic acid (ug/d) 14: Vit. B-12 (ug/d); 15: Retinol (ug/d); 16: Beta-carotene (ug/d).

As so many factors are difficult to be considered together, for practical purpose, general formulations are usully made by the nutritionist-cum-dietitians to design balanced diets keeping in view some selected factors that are of greater significance. Such type of formulations are done for a particular community, region, zone or for the people of a country as a whole.

The dietary regime which is considered as balanced diet that has been recommended by the Indian Council of Medical Research (1968) respectively for men, women, adolescents and children has been presented in Tables 18–21.

Table 18. Balanced Diet (Daily) for Men (gram per person)

Food Articles	*Mode of Work*					
	Heavy		*Moderate*		*Sedentary*	
	**V*	*NV*	*V*	*NV*	*V*	*NV*
Cereals/millets	650	650	475	475	400	400
Pulses	80	65	80	65	70	55
Leafy vegetables	125	125	125	125	100	100
Other vegetables	100	100	75	75	75	75
Tuber crops	100	100	100	100	75	75
Fruits	30	30	30	30	30	30
Fats and oils	50	45	45	40	40	35
Milk	300	100	300	100	300	100
Meat and fish	–	60	–	50	–	30
Egg	–	30	–	30	–	30
Sugar/jaggery	55	55	40	40	30	30

*V: Vegetarian; NV: Non-vegetarian.

N.B.: 50 gram of groundnut is recommended to those who do heavy physical work.

Table 19. Balanced Diet (Daily) for Women (gram per person)

Food Articles	*Mode of work*						*Extra Amount*	
	Heavy		*Moderate*		*Sedentary*			
	**V*	*NV*	*V*	*NV*	*V*	*NV*	*P*	*L*
Cereals/millets	475	475	350	350	300	300	50	100
Pulses	70	55	70	55	60	45	–	10
Leafy vegetables	125	125	125	125	125	125	25	25
Other vegetables	100	100	75	75	75	75	–	–
Tuber crops	100	100	75	75	50	50	–	–

Food Articles	Mode of work						Extra Amount	
	Heavy		Moderate		Sedentary			
	*V	NV	V	NV	V	NV	P	L
Fruits	30	30	30	30	30	30		
Fats and oils	50	45	40	35	35	30		
Milk	200	100	200	100	200	100	125	125
Meat and fish	–	50	–	30	–	30	–	–
Egg		30		40		30		
Sugar/jaggery	40	40	30	30	30	30	10	20

V: Vegetarian; NV: Non-vegetarian; P: Pregnancy; L: Lactation.

N.B. : 40 gram of groundnut is recommended to those who do heavy physical work.

Table 20. Balanced Diet (Daily) for Adolescents (gram per adolescent)

Food Articles	Boys of Years				Girls of Years	
	13 – 15		16 – 18		13 – 18	
	*V	NV	V	NV	V	NV
Cereals/millets	430	430	450	450	350	350
Pulses	70	50	70	50	70	50
Leafy vegetables	100	100	100	100	150	150
Other vegetables	75	75	75	75	75	75
Tuber crops	75	75	100	100	75	75
Fruits	30	30	30	30	30	30
Fats and oils	35	40	45	50	35	40
Milk	500	200	400	150	400	150
Meat and fish	–	60	–	70	–	80
Egg	–	30	–	30	–	30
Sugar/jaggery	55	55	40	40	30	30

Table 21. Balanced Diet (Daily) for Children (gram per person)

Food Articles	*Children of Years*							
	1 – 3		*4 – 6*		*7 – 9*		*10 – 12*	
	**V*	*NV*	*V*	*NV*	*V*	*NV*	*V*	*NV*
Cereals/millets	150	150	200	200	250	250	320	320
Pulses	50	40	60	50	70	60	70	60
Leafy vegetables	50	50	75	75	75	75	100	100
Other vegetables and tubers	30	30	50	50	50	50	75	75
Fruits	50	50	50	50	60	60	60	60
Fats and oils	20	20	25	25	30	30	35	35
Milk	500	200	400	200	400	200	400	200
Meat, fish and egg	–	30	–	30	–	40	–	50
Sugar/jaggery	30	30	40	40	50	50	50	50

V: Vegetarian; NV: Non-vegetarian.

N.B.: 50 gram of groundnut is recommended for boys of 16–18 years.

Chapter 10

Concluding Part

The late Dr. Amiya Kumar Bose, the eminent nutritionist, founder of the Indian Dietetic Association and the Professor of Obstetrics and Gynaecology, National Medical College in Kolkata quoted, "Malnutrition of people is not only a health issue but it is also a serious stumbling block for upliftment of human society and idealistic growth of human civilization". A great many biologists, sociologists and the great authorities on humanism have opined that malnutrition among people has been a pertinent factor for many evils of the present day.

To eradicate malnutrition among people in a country like India, plenty of factors associated with this repugnant food habits are needed to be heedfully paid attention to and efficiently controlled. Some of these aspects have been brought to light in the following:

(1) Knowledge on Nutrition

What really nutrition is and what are its benefits must be made known not only to gentlefolk and those belonging to the higher society but also to the common people, educational status of whom is not high. A considerable part of the population in India and also in many other countries is found to have the impression that malnutrition refers to less intake of nutritive food, that is, under-nutrition (= under-nourishment). In other words, they are not aware of what under-nutrition and over-nutrition are and the difference between the two. Hence, the ordinary common people should be made known of the fact that *both under-nutrition, i.e.,* intake of relatively less amount of the nutritive foods *as well as over-nutrition, i.e.,* bulking or over-feeding of nutritive food, both come under the purview of malnutrition.

What is meant by nutrients, what are the components of nutrients and how each of them benefits the body should also be made known to the people of all sections. It is also essential to make them aware of the harmful effects of any nutrient if it goes to the body in super-optimum level. It is often observed that some people are not aware of the fact that vitamin is not one and there are several types of vitamins. That each of the vitamin has specific role, functions and beneficial effect in the body is not known or clearly known to them. Hence, it becomes necessary to make the people aware of the *individual components of each nutrient.*

Imparting the knowledge on what foods are to be taken by an individual and their amount, in order that the body gets all the nutrients and to the proper amount cannot be set at defiance.

Misconception of nutrients is found among people in many other ways. Many women in the rural parts of India have the belief that cow's or buffalo's milk is the best food and being led by this belief, they feed the infants with cow's or buffalo's milk even without diluting with water. Many rural women consider ghee as the highly nutritive food and they serve high amount of pure ghee to the family members in their diet. There is report that in some parts of India, rural people abstain from taking common salt which has been iodized and even distributed to them without charging anything.

Knowledge on nutrition to the common people has its necessity through the audio-visual media. However, number of educators in the area on nutrition and dietetics is also needed to be increased to them by imparting formal training on extension education.

(2) Provision of Low-cost Food

"One of the major causes for malnutrition in India" has been stated to be, "economic inequality". For a part of the population in the country, low economic status stands in the way to secure food for desirable quality, nutritive value and also quantity. Consumption of meat, fish, egg and milk is much on the lower side by the poorer section due to high cost for which they are deprived of animal source protein and live on protein of plant sources only. A report states, "Inadequate protein is consumed by the Indian people because 56 per cent of poor Indian household consume cereals for protein requirement and the protein derived from cereals is inferior in quality than animal source protein".

Poultry keeping and goat husbandry need encouragement by the poorer section to secure meat, egg and milk. It may be stated that cost factor comes at the lower side for the culture of some fishes. Fish culture even in the paddy fields had also been a practice in India some decades ago but this is threatened now due to *indiscriminate use of chemical fertilizers and inorganic pesticides* to the crop plants. Paddy-cum-fish culture has many advantages like economical utilization of land, additional income, control of insect pests in the field *etc.* Some of the aquatic weeds are known to be controlled by some fishes like tilapia and the common carp and so on. To encourage this practice, organic farming of rice needs attention.

Researches undertaken by the co-author have revealed that in case of many fruits, vegetables, grains, spices, condiments *etc.*, the peels and the parts adjoining the peels have higher amount of nutrients than the inside part. (Ref. B.C. Mazumdar. *Wealth from Waste*, The Statesman Daily, Calcutta). In many fruits, sweetness is also relatively high in the inner than the outer part. Hence, cooking these foods without peeling where possible might ensure higher nutritive value of the cooked food.

In cooking rice, the water which is left after boiling is usually discarded by many Indian people. But it should be noted that the water left after boiling rice is a *rich source of starch besides other water-soluble vitamins, minerals, amino acids etc.* Hence, this water may be used to cook other foods and curries so as to make the products nutritionally enriched with no extra expenditure.

(In this connection, it may be also be mentioned that the water with which the rice is washed before boiling it with fresh water also contains many valuable substances. The water with which the rice is washed and then thrown away is a rich source of ferulic acid, allantonin, some vitamins, amino acids, minerals *etc.* Ferulic acid is a good antioxidant and when topically applied on skin protects it against ageing symptoms. Allantonin has moisturizing effect and nourishes skin when applied topically. It is an amniotic fluid, which is the end-product of purine metabolism and also quickens wound healing. Rice-washed water also contains inositol, which is hexa-dydroxy-cyclo-hexane and is known to stimulate hair growth when applied on scalp).

(3) Children's Nutrition

Special attention is necessary to be paid in India in regard to nutrition and diet of children. It is on record that more than one-third of the world's malnourished children live in India and half of the children among them under three years of age are in fact, underweight. On contrary, it may also be stated that one-third of the wealthiest children are overfed, which means that they are actually over-nourished. Thus, it may be opined that *malnutrition of the children in India is two-fold.* While under-nutrition is particularly apparent in the rural areas, over-nutrition is more common for the urban children and both need to be checked.

It is heartening that the Govt. of India started "midday meal scheme" on August 15, 1995 which aims at serving children with fresh cooked meals in a large number of schools in the country. Integrated Child Development Scheme, National Plan of Action for Children and other projects launched by the Govt. of India and the Children's Fund of the United Nations have made great strides to lessen the burden of under-nutrition of the children in India.

Over-nutrition of the children in the urban areas is to be discouraged by imparting proper education on nutrition to the urban families.

(4) Higher Production of Nutritive Food

In agricultural practice, growing varieties of crop plants which are known to produce yield of higher nutritive value needs encouragement. In the rural areas, setting up of "Nutrition Gardens" by the Panchayats, schools and other public and private institutions needs encouragement. These gardens may not be large in size but a wide variety of crops that are of high nutritive value may be grown in them to meet the requirement of the villagers.

(5) Using Variety of Foods

Attention should be paid that the diet of a person should consist of a variety of food items and *not food* of *one or two types only*. By the use of many types of foods, a wide variety of nutrients are possible to be taken up by an individual. The benefit of the traditional use of cereal-pulse mixture in the diet of Indian people has been stated in the Chapters 2 and 3.

(6) Use of Dietary Fibres

Dietary fibres and their health benefits have been stated under the Chapter on Carbohydrates (Heading 2.8 b). Although till now these foods are not considered under the category of nutrients but their enormous benefits on health have been well-founded. It is established now that every person barring the infants should take everyday certain amount of soluble and insoluble forms of dietary fibres. What type of foods and how much of them should be taken by a person daily to meet the requirement of dietary fibres should however, be recommended by the nutritionist-cum-dietitian keeping in view the health of the person and other factors.

(7) Exploring New Foods

With ever-increasing rate of population, agricultural land is getting squeezed instead of making increase. This obviously reflects in production of food in India that is not commensurate to meet the requirement of the people. Hence, exploration of other *non-conventional foods* is an aspect that deserves attention. Mention may be made in this connection that quite a number of plants that grow here and there mostly as weeds on lands and tanks in various parts of West Bengal are suitably cooked and consumed by the people of West Bengal as vegetables and greens. Many of these plants have been experimentally found to outdo many of the conventionally used agricultural crops in respect of nutrients and medicinal properties.

Seas and oceans are precious gifts of the Mother Nature. They cover about 71 per cent of the geographical area in the world and apart from storing of vast hidden treasure, they host innumerable species of flora and fauna to live in them. Many of these biotic agents are used as food by people living in many of the coastal countries since ancient time but attention on these food items could be said to be less by the Indian people.

The plants grown in the seas and oceans, which are popularly known as marine weeds or marine vegetables have great nutritional superiority particularly in respect of iodine, polysaccharides and many of the B vitamins. There are plentiful species and are usually recognized as red, green and brown coloured algae. Some of these algal species that may be used as food are dulse, carola, arame, hypnea, gim, chlorella, kombu, hiromi, oarweed, sargassum, hiziki, bladderwrack *etc*.

Sea animals that can be used as food are of diverse types. Among these, examples of some of the fishes are, demersal, diadromous, marine pelagic *etc.*, molluses are cephalopods, gastropods, crabs, shrimps, *etc.*, and among crustaceans, lobsters, krill *etc.* are common examples. Plentiful species of mammals and other animals also abound in the sea but their use as food is restricted. It should be remembered that adequate care and precautions are necessary to be adopted in regard to sea organisms, whether plants or animals as many of them may have toxic effect.

(8) Nutrition of Elderly Persons

Nutrition of the elderly people perhaps does not receive adequate attention in India. It should be noted that both under-nutrition and over-nutrition seriously act upon the health of the elderly people in many ways. Some of these may be stated as fatigue, weakness of muscles, constipation, lack of good sleep, loss of weight, memory loss, weak immune system, mental depression, hypertension, various types of phobia, breathing difficulty, osteoporosis, difficulty to hold urine, insecured feeling *etc.*

Proper nutrition and diet have the potential to *overcome many of the health problems of old age*. This has been discussed under the heading, 9.12 in the Chapter on Dietetics. The diet charts provided by Pasricha, S. and Thimayamma, B.V.S., *Dietary Tips for the Elderly*, 2004, published by the National Institute of Nutrition (I.C.M.R.), Hyderabad - 500 007 may be of great help for the elderly persons.

(9) Cooking

Cooking of food has now received the status of culinary science and technology. It has a great impact not only on the palatability of the foods but also on their nutritional value. It has been stated under the Chapter 5 that activity of some vitamins are lost when cooked at high temperature or if cooked in uncovered condition. Some pulse grains are difficult to be digested due to the tough proteins that are present in them. Pulses become easily digestible and cooking can be done in short time if soaked in water to become swelled (= imbided)

Use of sprouted pulse grains, *i.e.*, making them germinated to protrude the white radicles (= primary roots) and then cooking them is of high nutritional benefit. This is because, the protein content in the grains is *hydrolyzed to liberate free amino acids* which are digested and absorbed more easily than the protein present in the unsprouted pulse grains. Vitamin-C and some of the vitamins are also increased in the grains when they are sprouted. This can be done easily by pre-soaking the grains overnight and then maintaining within occasionally moistened cloth for one or two days under room temperature condition.

Spices and condiments are excellent substances which bring about savourishness of cooked foods. However, spices and condiments of various types have also great food value apart from their tremendous role in digestion of food. It may be mentioned in this connection that tamarind is much used in cooking of food. Tamarind is highly valued for its digestive capacity and nutritive value. Its role to prevent heart disease is also claimed but needs adequate experimental support. Use of coconut milk and ground coconut in cooking food is a common practice in the southern and the eastern India. These ingredients also add nutritive value to the cooked food, apart from enhancing satiety.

(10) Danger of Chemical Contamination

Chemical pollution of foods has become an alarming situation at the present time which not only robs the quality and nutritive value of them but also causes serious health hazards. Indiscriminate use of chemical insecticides on agricultural crops, such as Endosulfan, Propoxur, Chlorpyriphos, Carbaryl *etc.* is a causative factor for nervous disorders, liver complications, thyroid abnormalities, cardiovascular diseases and even cancer when the crops treated with these chemicals are consumed. Entry of insecticides, *viz.*, Dichlorvos, DDT, Carbofuram, Malathion *etc.* into the body through food may cause paralysis, nervous disorders, eye troubles and also cancer.

Consumption of crops treated with the fungicides, *e.g.*, Ziram, Carbendazim, Thiabendazol *etc.* may cause liver complications, Parkinson's disease, eye troubles *etc.* Treatment of crops with weedicides, *e.g.*, Paraquat, Triazine, 2, 4-D, Glyphosate *etc.* may bring about Alzheimer's disease, thyroid tumour *etc.* on consumption of the treated crops. Fruits artificially ripened on exposure to acetylene gas causes serious health hazards, particularly to the pregnant women. Serious harm is also produced in the body on consumption of foods treated with synthetic colours, *e.g.*, Red - 40, Citrus Red - 2, Yellow - 5, Indigo carmine, Red - 3 *etc.* To get rid of such health hazards due to entry of such toxic chemicals into the body through food, *organic method of cultivation of agricultural crops* is warrantable.

Appendix I

Glossary

Abdomen – Body part between pelvis and thorax which contains stomach, lower part of oesophagus, liver, intestines, gallbladder, spleen, pancreas and urinary bladder.

Abdominal crisis – Severe pain in abdomen, which usually occurs in sickle cell anaemia or syphilis.

Absorbefacient – Any substance which causes absorption.

Acetone – Dimethyl ketone. Highly volatile, colourless, inflammable, organic solvent of sweet smell. Occurs in blood and urine during diabetes and some other metabolic disorders.

Acidolysis – Chemical reaction which involves decomposition of a molecule by addition of the elements of an acid.

Acidosis – Increase of acidity in blood which may be due to accumulation of acid (in diabetes or kidney disease) or heavy loss of bicarbonate (in kidney disease).

Activation – Increasing activity of one material by addition of another.

Acyl – Carbonyl group which is attached to the alkyl or aryl group.

Adenosine - Nucleotide containing adenine and ribose.

Adsorption – Attachment (physical) of gas or liquid molecules on (usually) a solid surface.

Adulteration – Addition or substitution of impure or harmful substance in a genuine product.

Agar – Dried form of a mucilagenous product obtained from some sea algal species.

Albuminuria – Presence of serum protein, *e.g.*, albumin, globulin *etc.* in urine.

Alimentotherapy – Treatment of disease by regulating diet (Syn. Dietotherapy).

Alkalosis – Increase of blood alkalinity by accumulation of alkali or lowered level of acid.

Allergen – Agent (antigen) which causes allergy and is of diverse types.

Amnesia – Hysterical episodes of forgetfulness of one's own identity.

Angiotensin – converting enzyme inhibitor – Agent which inhibits conversion of Angiotensin I to II and causes vasodilatation (dilatation of arteries and arterioles) and high flow of renal (kidney) blood.

Antidote – Substance which neutralizes a poison. It may be chemical, physiological, mechanical or universal.

Arteriosclerosis – Thickening, hardening and loss of elasticity in any part of arterial wall.

Atherosclerosis – Narrowing of artery due to deposition (plaque) of fat, cholesterol and calcium. Atherosclerosis is a type of arteriosclerosis and the term should not be interchangeably used for arteriosclerosis.

Basal metabolic rate – Metabolic rate which is measured under some basal condition. *e.g.*, after some hours of eating, after sleep *etc.* and is expressed as kilocalories per square metre of the surface of body per hour.

Benign – Non-malignant and not recurrent.

Bile – Bitter and think fhick of the liver, which passes from liver into the duodenum by bile duct.

Blood count – Number of red corpuscles and luekocytes per microlitre of blood.

Blood pressure – Pressure exerted by the flowing blood on arterial walls.

Blush – Reddish patch on face and neck due to vasodilation.

Bolus – Food which is to be swallowed after mastication.

Bond – Force which binds ions or mass.

Bone marrow – Soft material (organic) which fills bone cavities.

Bradycardia – Slow heartbeat having a pulse rate of usually less than 60 beats per minute in an adult.

Buccal cavity (Syn. Bucco-pharyngeal cavity) – Mouth cavity.

Calcination – Drying and grinding.

Calcinosis – High deposition of lime salts in tissues.

Calcipenia – Deficiency of calcium in tissues.

Calcitonin – Hormone product by thyroid gland which maintains strong bone.

Calorie – Unit of heat defined as heat which is required to raise temperature of 1 gram of water from 14.5°C to 15.5°C.

Cancer – Malignancy which results in uncontrolled growth of cells.

Canker sore – Ulcer of lips and mouth cavity.

Capillary – Blood vessels which are very minute and connect smallest arteries with smallest veins.

Caput – Head or extreme part of an organ.

Cataract – Eye abnormality resulting opacity of the lens or capsules.

Cation – Positively charged ion which travels to the negative pole, *i.e.*, cathode in solution.

Cereal – Grains of glass family which are edible. Major cereals are rice, wheat, maize and barley.

Chondrology – Science of cartilage.

Co-enzyme – Co-factor of enzyme and often, a vitamin.

Colon – Large intestine from end of ileum to the anal canal.

Collagen – Insoluble, fibrous protein of connective tissue, such as dermis, ligaments, tendons *etc.*

Decomposition – Breaking down of an organic form to liberate simpler forms.

Dehydration – Excessive loss of water in a tissue or organ.

Dementia – Loss of memory, a cognitive deficit.

Depressant – Any agent which decreases nerve activity.

Detoxification – Removal of toxic properties of a poison.

Diphosphate – Containing two phosphoric acid groups.

Diuretic – An agent which increases secretion of urine (used to lower high blood pressure).

Diverticulum – A small pouch in the wall of colon.

Dysuria – Difficult urination with pain.

Ebullism – Water vapour formation in tissues which results in exposing body in reduced barometric pressure.

Emaciation – Becoming extremely lean due to wasting disease.

Embolism – Stoppage of blood flow in artery or vein due to some foreign substance or blood clot.

Emesis – Vomiting to expel contents of stomach.

Endocrine gland – A gland which was no duct and the secretion is directly discharged into blood or lymph.

Enema – Introduction of water or a solution into the rectum to cause easy defaecation.

Enteritis – Inflammation of the intestines usually of mucosa and submucosa.

Epidemic – A disease which is infectious and attacks many people at the same time in a locality.

Erithropoiesis – Formation of red blood cells.

Evacuation – Emptying of bowels.

Exhalation – Breathing out process.

Exocrine gland – A gland whose secretion comes to epithelial surface through a duct or directly.

Expiration – Removal of air from lungs by breathing out.

Faeces – Waste material excreted from bowels through anus.

Fissure – A cleft or deep furrow in any organ in the body.

Flatus – Expulsion of gas from the intestine through anal passage. A person may expel upto 1200 cc of gas in 24 hours.

Flavin adenine dinucleotide (FAD) – Co-enzyme in some oxidation-reduction enzymes and contains flavin.

Fluid retention – Inability to expel fluid from the body due to renal, cardiac or other umpairment.

Flux – Excessive discharge from an organ of the body.

Follicle – A cavity or sac which is small.

Food preservative – Any substance which preserves food against microbial contamination and chemical spoilage.

Fungicide – Any substance which kills fungi and fungal spores.

Gall – Bitter secretion of liver stored in gall bladder.

Ganglion – Mass of nervous tissue containing mainly neuron cell bodies.

Gastric lavage – Washing of stomach to expel out the contents.

Gastrin – A hormone which is secreted by the mucosa of the pyloric area of stomach and duodenum.

Genitalia – Reproductive organs.

Geriatrics – Medical speciality concerned with ageing. Also known as gerontology.

Germ – A micro-organism which causes disease.

Gestation – Period from conception to birth.

Gestosis – Any disorder occurring during pregnancy.

Gingivitis – Inflammation of gums.

Glucagonoma – Malignant tumour of the alpha cells of the islets of Langerhans.

Habitus – Physical appearane or attitude.

Haemoglobin – Pigment of the red blood cells which contains iron.

Helminthic – Property of expelling worms from the body.

Hepatitis – Inflammation of the liver.

Hernia – Protrusion of an organ or part of an organ through the wall of the cavity which contains it.

Hiccup – Spasmodic closure of the glottis that occurs periodically.

Human immunodeficiency virus (HIV) – Retrovirus that cause acquired immunodeficiency syndrome (AIDS).

Hydration – Chemical combination with water.

Hydrion – Hydrogen ion.

Hydropsy – Oedema.

Hygroma – A sac which contains fluid.

Hyperemia – Accumulation of blood in a part of the body.

Hypoacidity – Decreased acidity in stomach.

Ideation – Thinking to form ideas, which reduces maniac depression and brain diseases.

Ileum – Lower part of small intestine from jejunum to ileocecal valve.

Imbibition – Absorption of water or fluid by cells or tissues.

Immunity – Resistance against disease by a pathogen or due to development of antibody.

Immunization – Producing immunity against a disease.

Immunodeficiency – Decreased ability to respond to antigenic stimulus.

Immunodeficiency disease – Autosomal disorder due to some dysfunction to immune system.

Incision – Cutting an organ with a knife.

Infarct – A part of an organ in which blood supply is totally lost and hence, death of that part (necrosis) occurs.

Infection – invasion and growth of micro-organisms in a tissue resulting damage.

Inflammation – Reactive process in a tissue or an organ due to injury which is caused by thermal, electrical, mechanical, chemical or due to some infective agent. Inflammation is named by suffixing the word, itis with the organ or tissue, which is inflammed.

Instinct - Phenomenon to react to some environmental conditions and stimuli in some characteristic manner, which is inherent tendency of animals and human beings.

Intestine - Alimentary canal from pylorus to anus. It has two parts which are small intestine and large intestine or colon.

Intracellular - Within the cell.

Intrauterine - Inside the uterus.

Intrinsic - Any structure belonging solely to some body part, *e.g.*, intrinsic muscles.

Irrigation - Cleaning of a body part by washing with water or other fluid.

Jejunum - Second part of small intestine from duodenum to ileum.

Joint - Articulation and juncture between two adjacent bones.

Joule - Heat energy required to move 1 kg mass 1 metre by 1 Newton. (1 Cal = 4.184 joule)

Karaya gum - Gum derived from plants of the family, *Sterculiaceae*, which is dried and used as a bulk laxative.

Kenny treatment - Physiotherapy to treat poliomyetitis.

Keratin - Highly tough protein present in nails, hair and horny tissue. May be harder or softer.

Keratitis - Corneal inflamation with decrease vision.

Lactation - Secretion milk and sucking period in mammals.

Lactose intolerance - Intolerane to milk and other foods containing lactose. It causes gastrointestinal disorders. This may be an inherent character or acquired later, after several years of ageing.

Lactosuria - Occurrence of lactose in urine, which usually occurs in pregnancy and lactation.

Laparotomy - Opening of abdomen by surgery.

Lard - Purified fat of hog.

Laryngorrea - Very high discharge of laryngeal mucus.

Laser - Acronym for the words :

Light Amplification by *Stimulated Emission* of *Radiation*. Laser focussed through instrument is treated for many diseases, *e.g.*, retinal detachment, diabetic retinopathy, removal of warts, skin cancer *etc.*

Laxative - Medicine which looses bowels and increases peristalsis.

Lymph - Clear, transparent, alkaline fluid in the lymphatic vessel.

Macroscopy - Examination of a part with naked eye.

Malaria – Protozoa (Plasmodium) attacked disease in red blood cells. (Formerly considered wrongly as due to mal air). It was wrongly believed that mal (= bad) air as the cause and hence, the name, malaria was proposed.

Malignancy – Tumour or neoplasm which is cancerous.

Malt – Germinated grain mostly of barley and used as food in wasting disease.

Maltosuria – Excretion of maltose through urine.

Mammary glands – Glands of female breast which secrete milk.

Mania – Mental disorder with excess excitement.

Manual – Performed by hands.

Mastication – Chewing.

McArdle's disease – Abnormality when glycogen is absorbed in muscles in very high amount, which is due to deficiency of myophosphorylase.

Medicament – Medicine.

Melanoma – Dark, malignant tumour of skin.

Mentation – Mental acivity. (Syn. Mind-set).

Metria – Ulterine inflammation during pregnancy.

Microbiota – Micro-organisms in an area.

Microsmatic – Poor sense of smell.

Milk of magnesia – Magnesium hydroxide in suspension used as antacid which looks like milk.

Monophosphate – Cyclic form of adenoside.

Motility – Having power to move at will.

Mucin – Glycoprotein which is present in mucus.

Mucosa – Mucous membrane that lines hollow organs.

Mycosis – Disease caused by fungus.

Necrosis – Death of some tissue or bone but surrounded by healthy tissues.

Neonate – Newborn infant upto one year of age.

Nephrectomy – Removal of kidney by surgery.

Nephritis – Structural and functional unit of kidney which contains renal capsule.

Nerve fibre – Elongated neuron.

Neurolysis – Nerve stretching to alleviate pain.

Nuromyelitis – Inflammation of nerve and spinal cord.

Neuropacemaker – Electrical stimulation of spinal cord by device which is implanted in the body.

Neuropathy – Any nerve disease.

Nissl body – Granular and large body in nerve cell.

Normocyte – Red blood cell which is average in size.

Nutation – Nodding of head involuntarily.

Obesity – Abnormal deposition of fat (corpulence) in the body.

Occiput – Back (rear) part of skull.

Occult blood – Minute bood that comes through faeces or from other organs, apparently not noticeable.

Ointment – Antiseptic greasy substance used for application on injured skin.

Onychomycosis – Fungal infection of nails.

Ophthalmalgia – Pain in part of eye.

Ophthalmia – Inflammation of eye, usually when serious.

Opthalmologist – Eye physician or surgeon.

Optometrist – Refractionist who recommends power of lenses but does not treat disease of eye which is done by an ophthalmologist.

Oral rehydration – Administration of electrolytic solution through mouth to protect against dehydration.

Osteitis – Inflammation of a bone.

Osteoporosis – Reduction in bone density.

Otitis – Inflammation of ear.

Oxygenation – Addition of oxygen in blood (Not oxidation).

Oxygen toxicity – Various high and prolonged oxygenation in bood which causes respiratory failure.

Pacemaker – In heart disease, this device is implanted (cell or group of cells) such that it generates impulses automatically.

Palpebra – Eyelid.

Palsy – Paralysis.

Palynology – Study of pollen grains of the anthers of flowers having importance in study of allergic reactions.

Panacea – An agent that remedies any type of illness, which though is practically impossible.

Pancreatalgia – Pain in pancreas.

Pancreatectomy – Surgical removal of part on whole of pancreas. (Ectomy = surgical removal).

Papilla – Nipple-like, small growth or protuberance.

Papilloma – Epithelial, non-malignant tumour.

Papillomavirus – Some viruses which cause papillomas on the body surface.

Paradox – Something which apparently be true but practically not so.

Paralysis – Loss of sensation of a body part, which may be parmanent or temporary and is an abnormality of motor neurons.

Paraprotein – Abnormal protein in plasma, *e.g.*, macroglobulin.

Parasite – An organism which invades an organic body and derives nourishment from it.

Parkinson's disease – Disease of the nervous system which causes tremor, weakness of muscles, gait and paralysis.

Parturition – Childbirth.

Pasteurization – Destorying part of the microbes in a liquid food by heating but not boiling and followed by immediate cooling.

Pathogen – A micro-organism or a substance which when invades the body part causes diseases.

Pediatrics – Medical speciality which deals with diseases of children.

Pedicure – Care of the feet, particularly for cosmetic purpose.

Peristalsis – Involuntary, wave-like movement occuring in hollow part of the body, *e.g.*, alimentary canal.

Phosphorism – Poisoning from excess phosphorus.

Plexus – Nerves, blood or lymphatic vessels in network.

Protein-C – A protein in blood plasma which prevents excess clotting. Thrombosis results when it is deficient.

Proteinuria – Excretion of protein, mostly albumin throgh urine.

Pyrosis – Heartburn. (This is not heart disease).

Quadriplegia – Paralysis caused by spinal cord injury.

Quick's test – A type of liver functioning test.

Radiation – Ionizing ray emitted from a source and is used for diagnosis or as a therapentic agent.

Rash – Eruption on skin.

Raving – Talking as in delirium.

Refractionist – Eye specialist who provides power of lenses in visionary defect. Optometrist.

Regimen – Regulation of diet, a sleep and exercise in a person which may benefit health.

Regurgitation – Flow of the contents of stomach to the mouth.

Renal transplantation – Implantation of kidney donated by another person.

Reovirus – A virus which inhabits in respiratory or digestive systems of healthy persons.

Referfusion – Blood supply in heart muscle which is ischaemic.

Resistance – Resisting invasion of any pathogen by the body.

Resorcinol – Any chemical which has kerotolytic and cidal effect to micro-organisms and used in skin abnormalities.

Resorption – Removal of pus or toxic substance by absorption.

Respirator – Device for helping in respiration.

Resuscitation – Forcing air into the lungs from the mouth of a person or any device which is done to the patient in starvation of oxygen.

Retina – Layer of the eyeball which is in the innermost part and receives image of the object through lens.

Retinopathy – Any type of disorder of the retina, which may be due to diabetes, high blood pressure *etc.*

Rhinopathy – Nasal disease. (Rhinitis).

Ribs – Narrow curved bones of 12 pairs in breast.

Rind – Outer coat of an organ.

Rods and cones – Retinal cells which are phtoreceptor.

Rosacca – Disease of the facial skin.

Sac – Small pouch.

Saccharin – Artificial sweetener, prepared from coal-tar.

Saline – Containing salt. Hypertonic saline solution contains salt higher than 0.85 per cent and in hypotonic saline solution, salt concentration is less than 0.85 per cent.

Salivant – Agent which increses flow of saliva.

Salt – Common edible salt which is sodium chloride.

Sanitation – Creating condition favourable to heatlh by adoption of suitable measure.

Saprophyte – An organism which inhabits on dead or decaying organic matter.

Sarcology – Medical speciality which deals with relatively soft tissues of the body.

Sarcoma – Malignancy of muscle or bones (connective tissues).

Scabies – Skin disease caused by an arachnid of arthropod.

Scapula – Large, triangular bone at the posterior part of shoulder.

Schizophrenia – A psychic disorder which affects thinking, attitude, perception, behavour, attention, delusion, hallucination *etc.*

Sciatic nurve – Nerve which arises from sacral plexus on each side and largest in the body.

Sclerosis – Hardening of a tissue, organ, blood vessel *etc.*

Sebum – Fatty substance of the sebaceous gland of the skin.

Sedentary – An occupation in which very little physical exercise in done.

Senium – Old age.

Serum – Watery fluid matter of blood after coagulation.

Sorbitol – An alcohol which is crystalline and is present in many fruits.

Spasm – Sudden muscular contraction that occurs due to some disturbing and irritational substance.

Spinal column – Vertebral column.

Steroid – Organic compound, whose nucleus contains perhydro-cyclo-pentano-phenanthrene.

Substrate – A substance on which enzyme acts.

Suppuration – Pus formation.

Symptom – Change in the body parts which is perceptible, indicating disease. Symptoms may be objective, subjective or cardinal.

Syndrome – Several symptoms and signs of disorder which function together.

Tachycardia – Rapid heart action, which may be more than 100 beats per minute in an adult.

Tannin – Acidic substance present in bark of some plants and is used as an astringent or antidote of poisons.

Tendon – Connective tissue which attaches muscle to bones.

Tetanus – Infectious disease of the central nervous sysem which is caused by endotoxin of tetanus bacillus.

Tetany – Nervous abnormality which causes nervousness, irritability, numbness, cramps *etc.*

Therapist – One having skill in therapy.

Tincture – Alcoholic extract of plant or animal source material.

Tinea – Skin disease, ringworm.

Tolerance – Ability to endure any drug, food, chemical, climate *etc.*

Toluene – Hydrocarbon derived from coal-tar.

Tonic – Medicine which tones up body.

Tonsil – Lymphatic tissue in the mucous membrane of fauces and pharynx.

Toxicopathy – Disease caused by a posionous substance.

Tranquilizer – Any medicine which alleviates mental tension but does not affect mental acivity.

Trauma – Wound or injury.

Tremor – Involunary movement of body part.

Trichobezoar – Hairfall.

Trichology – Study of hair and scalp and tratment of diseases of those parts.

Trichosis – Any disease of hair.

Triphosphate – Containing three phosphoric acid groups.

Turpentine – Mixture of terpene and hydrocarbons obtained from pine trees.

Ulcer – Sore of the skin.

Universal antidote – An antidote which is used against any type of poisoning, and contains activated charcoal, tannic acid and other materials.

Vaccination – Vaccine inoculation in order to prevent infection of specific pathogen.

Varicose vein – Highly enlarged surface vein which is painful.

Vasectomy – Surgical removal of the whole or part of vas deferens.

Vasodilation – Swelling of small arteries and veins.

Vector – Carrier wich transmits disease organisms.

Virus – Organism which is smallest and lives inside cells.

Wool fat – Lanolin. Only fat which absorbs water. Used in making ointment.

Wrinkle – Ride in the skin.

Xanthoderma – Yellowing of skin.

Xerosis – Excessive dryness of skin, conjunctiva or mucus membrane.

Yawn – Involunatary gaping to take deep breath.

Yeast – Saccharomyces fungi, which are unicellular and multiply by budding.

Yoghurt – Milk soured by bacteria which has therapeutic property and similar to curd.

Zoster – A type of harpes (nerve) disease.

Zygomatic bone – Bone of the face below eye.

Zymosterol – Sterol derived from yeast and has therapeutic value.

Appendix II

Units and Conversion

Weight

1 kilogram	=	1000 gram = 0.1 quintal
	=	0.001 tonne = 2.205 pound
1 quintal	=	100 kilogram
1 tonne	=	10 quintal = 100 kilogram
1 gram	=	0.001 kilogram = 1000 milligram
	=	106 mcg = 109 nanogram
	=	0.353 ounce
1 mg	=	0.001 gram = 1000 microgram
1 microgram	=	0.001 milligram
1 pound	=	16 ounce = 453.6 gram
	=	128 dram (drachm)
1 ounce	=	28.35 gram = 8 dram
1 carat	=	0.2 gram = 200 milligram

Volume

1 litre	=	1000 ml = 0.001 cu.metre
	=	0.264 gallon = 1.76 pint
1 millilitre	=	0.001 l = 1000 microlitre
	=	1000 nanolitre
1 microlitre	=	0.001 millilitre
1 fl. dr	=	60 minims
1 fl. ounce	=	8 fl. dram
1 pint	=	0.57 litre = 20 fluid ounce
1 gallon	=	8 pint = 3.785 litre

Length

1 metre	=	100 cm = 1000 mm
	=	3.281 feet = 39.37 inch
1 kilometre	=	1000 metre = 1093.6 yard
	=	0.62 mile
1 centimetre	=	0.39 inch
1 mile	=	1.61 km = 1760 yard
	=	8 furlong = 5280 feet
1 foot	=	0.3048 metre
1 yard	=	3 feet = 36 inch
	=	0.9144 metre
1 inch	=	2.54 centimetre

Index

D

E

F

N

O

P

R

S

T

U

V

W

X

Z

www.ingramcontent.com/pod-product-compliance
Ingram Content Group UK Ltd.
Pitfield, Milton Keynes, MK11 3LW, UK
UKHW021450280726
14060UKWH00001BA/340

9 789359 198538